HANDBOOK
OF
FOOD MANUFACTURE

FORMULAS AND TESTING

BY

DR. F. FIENE

Food Research Chemist

AND

SAUL BLUMENTHAL, B.S.

Consulting Food Chemist and Director of

SHIRLEY LABORATORIES

NEW YORK CITY

ABRIDGED EDITION

CHEMICAL PUBLISHING CO. OF N. Y., INC.

NEW YORK, N. Y.

PRINTED IN U.S.A.

Handbook of Food Manufacture

ISBN: 978-0-8206-0153-3

Chemical Publishing Company:
www.chemical-publishing.com
www.chemicalpublishing.net

First Edition: Chemical Publishing New York 1938

Printed in the United States of America

PREFACE

While there is no lack of good and authoritative books on foods, food chemistry, and the manufacturing of foods, there has not been published any simple book devoted to the many practical aspects of the food industry. Hundreds of requests have come to us, as consultants, from teachers, executives, buyers, brokers, foremen, and others in the industry who have not had the benefit of technical training.

Confident that "experience is the best teacher," we have selected from our experience the enclosed formulae and technique, which we know will prove most helpful. Theories and personal opinions have not been included.

All formulae given herein are being used or have been used on a large scale in this country and abroad. Some of these formulae have never been disclosed before. If directions are carefully followed good commercial products will result.

For the first time there is presented, in one book, grades and specifications of the numerous raw materials used in making the various types of food products.

The careful study of this book will prevent many spoiled batches, produce new and better products, and further a general understanding among those who do not have the time or access to the usual textbooks on the subject of foods.

We acknowledge most respectfully and with pleasure the valuable cooperation of our many friends in the industry who have constructively criticized our book in manuscript form. Other corrections and suggestions are invited and will be received most gratefully.

Dr. F. Fiene,

Saul Blumenthal, B.S.

New York City, N. Y.

CONTENTS.

CHAPTER I

MILK AND MILK PRODUCTS

Cow's milk is a white, opaque liquid having a butter-like odor and a bland sweet taste. Its specific gravity is about 1.030. Viewed under the microscope it is seen to be a clear and transparent liquid in which a large number of minute globules are suspended. On standing, these rise to the surface as a yellowish stratum, the cream, which constitutes about 5% of the milk and consists mainly of fat mixed with some casein and retaining some serum. By churning, the globules unite to form butter and the liquid which separates and is known as buttermilk is essentially a solution of milk sugar and the saline constituents of milk, and contains some casein and butter. The milk from which the cream has been separated is called skimmed milk. If left to itself, milk becomes acid and separates the casein in a coagulated condition called curds. The same effect is produced by Rennet and by acids. The liquid separated from the coagulum is called whey and contains chiefly milk sugar and some salts. From the curd, by different processes, cheese is made. The souring of milk is due to lactic acid bacteria, which enter the milk from the air, and transform the milk sugar into Lactic Acid ($C_3H_6O_3$).

DRY MILK POWDER

13	oz. milk powder	Make 1 gal. skim milk
7½	pt. water	

9	lb. butter	Make 4½ gal. (37 lb.) of 20% cream
2½	lb. milk powder	
3	gal. ½ pt. water	

1 gal. 4% milk weighs 8.5 lb.
1 gal. skimmed milk weighs 8.636 lb.
1 gal. water weighs 8⅓ lb.
1 gal. 20% butter fat cream weighs 8.4 lb.

STERILIZATION OF MILK

If fresh milk is boiled for 15 minutes a scum is formed, due to the coagulation of lactalbumin and globulin. This kills the bacteria which cause lactic fermentation, and the milk will remain sweet for several days if the receptacle is stoppered with cotton. Boiled milk is less digestible and nourishing as the fats, sugars, casein, and albumin are altered. When milk is heated at a temperature of 68°–75° C. (154°–167° F.) the taste and digestibility are not materially altered. A temperature of 68° C. is sufficient for the destruction of tubercle bacilli, 60° C. for typhoid, 58° C. for diphtheria, and most saprophytes will be killed at a temperature of 75° C.

PASTEURIZATION

This method is preferred to complete sterilization as it does not affect materially the nutritive value of milk. Pathogenic bacteria and non-pathogenic micro-organisms that cause the souring of milk are destroyed. Spores are not destroyed by this process.

Place the fresh milk in sterile glass bottles, stopper with cotton, and stand in a vessel of water (a double boiler), heat the water to 70° C. for ½ hour. The milk will remain sweet 24 hours.

"ACIDOPHILUS" MILK

The manufacture of pure acidophilus milk requires the greatest care, both in the preparation of the milk before inoculation and throughout the entire process of manufacture. It must be remembered that L. acidophilus is not a natural inhabitant of milk and must be acclimated to this medium. Even though it be acclimated to milk, its growth is nevertheless such that it is unable to overcome a slight contamination. Laboratory facilities, close supervision, and adequate equipment are essential to the successful production of acidophilus milk.

The first step in the manufacture of acidophilus milk is to obtain from a reliable source a pure culture of the bacterium Lactobacillus acidophilus of proved therapeutic value. After a pure culture is obtained it must be kept pure as a starter. All chances of contamination must be avoided. Florence or Erlenmeyer flasks are the most suitable containers in which to carry the starter. The flasks should be about half filled with fresh skim milk and plugged with cotton. They are then sterilized in an autoclave at 15 pounds pressure for 20 minutes. The milk after sterilization has a slightly caramelized appearance. After the milk has cooled, a flame from a gas burner should be passed over and around the mouth of the flask and the culture introduced into the flask after the mouth of the culture tube is passed through the flame. The cotton plug from the flask should be held in the hand during this operation and replaced in the mouth of the flask immediately after inoculation. The inoculated milk is incubated at 98° to 100° F. until the milk has curdled. A starter should show a small quantity of whey on top of the curd which should appear firm with no evidence of gas.

The curd is broken up by rotating the flask vigorously. The starter should be of a creamy consistency and possess a clean acid flavor and a characteristic aroma. This starter, or mother starter, is carried on by daily transfers from flask to flask. The same procedure as described above is followed except that usually the first bit poured over the lip of the flask is discarded as it serves to wash off the lip. The care necessary to maintain a pure starter of L. acidophilus cannot be overemphasized. The starter should be examined frequently for purity either microscopically or by plates. Should it develop a bad flavor or odor, it must be discarded and a new culture obtained. In any case it is advisable to get a new culture at regular intervals. Kopeloff advises the use of a new fecal culture every two to four months to assure the therapeutic value of the product.

The bulk starter, which is the starter to be used to inoculate the batch, is prepared in the same way as the mother starter; and the same precautions against contamination must be observed. Florence or Erlenmeyer flasks of 2-liter capacity or larger are most satisfactory in which to prepare this bulk starter. These flasks permit the autoclaving of the milk and are most easily handled under the conditions necessary to prevent contamination. The starter should be 18 to 24 hours old when used to inoculate the batch and should possess all the qualities indicating purity previously mentioned for the mother starter.

Equally as great care must be exercised in the treatment of the milk as in the preparation of the starter. Only fresh milk of the best quality should be used in the manufacture of acidophilus milk. Whole milk, milk which is partly skimmed, or skim milk may be used. The pasteurizer employed in the preparation of the product must be so fitted that the milk may be heated and cooled without removing it from the vat, which should be equipped with an efficient mechanical stirrer. The vat must be used not only as a pasteurizer but also as an incubator for the milk. The characteristics of the bacterium Lactobacillus acidophilus make it necessary that the bacteria in the milk be killed before the starter is added. The milk is heated to about 205° F. for an hour and a half or longer and then cooled to 98°. It should have a slightly caramelized appearance after heating; however, this in itself is not an indication that the necessary destruction of the bacteria has been accomplished. Bass recommends heating the milk to 190° to 195° for one hour; then cooling it to 98°, at which temperature it is held for three or four hours; again heating it to 190° to 195° for one hour; then cooling it to 98°; and inoculating. After the milk has been cooled to 98° to 100° it is inoculated with about 2% of the bulk starter. The greatest care must be observed in the inoculating process to prevent the contamination of the batch. The mouths of the flasks should be passed through a flame before pouring the starter. The cover of the vat is lifted just sufficiently to permit the pouring of the starter and is replaced immediately. After inoculation the milk is stirred for a few minutes to distribute the inoculum, and the batch is then allowed to incubate. The temperature of incubation, 98° to 100°, should be maintained throughout the entire incubation period. The batch is incubated until a firm curd has been formed, which usually requires about 18 to 24 hours. The length of the incubation period is dependent upon the size of the inoculation, the temperature of incubation, and the activity of the starter. After the milk has curdled it is broken up by stirring, cooled to room temperature, and bottled.

KEFIR

The yeast is prepared by adding a half teaspoonful of sugar to a 6-ounce or 8-ounce bottle of boiled and cooled water. Half a yeast cake is added to this sugar solution and set in a warm place overnight. This will give an active culture of the yeast and obviate the necessity for adding the yeast cake directly to the milk. This yeast culture should be ready at the time the buttermilk is received or, if made at home, at the time it is curdled.

One to one and a half per cent of sugar is added to the buttermilk. On the quantity of sugar added to the buttermilk will depend the extent of the alcoholic fermentation. Theoretically about one-half of fermented sugar may be converted into alcohol; that is, milk to which 1% of cane sugar has been added may contain after the fermentation 0.5% of alcohol. The quantity of sugar added should be governed by the amount of carbon dioxide it is desired to have in the finished product. This should be sufficient to make the kefir distinctly effervescent and impart to it the peculiar, sharp taste of charged water, but should not be developed enough to blow the fluid out of the bottles when the stoppers are removed. Experience shows that 1% to 1.5% of sugar will give the proper amount of gas. This may be approximated by adding sugar in the proportion of 2 even teaspoonfuls of sugar to each pint of milk. When the buttermilk and the yeast culture are ready, the sugar is dissolved in the buttermilk.

The yeast culture is added to the buttermilk in the proportion of 1 teaspoonful to 1 quart of buttermilk.

The buttermilk is mixed thoroughly and bottled. The bottles should be very strong, as sufficient gas pressure is sometimes generated to break ordinary bottles. The heavy bottles used for ginger ale or other carbonated drinks answer this purpose very well. They should be carefully cleaned and boiled or steamed before being filled and then stoppered tightly. The stoppers should be wired or tied securely in place.

The product is put in a cool place to ferment. If the fermentation is too active the kefir will have a yeasty taste, and the curd is likely to become lumpy and filled with large gas bubbles. A temperature of 18° to 21° C. (65° to 70° F.) will be found satisfactory for kefir which is to be used on the third or fourth day. The floor of a cool cellar is a convenient place to ferment kefir made in the home. The bottles should be shaken as often as may be necessary to keep the curd in a finely divided condition. The finished product should be smooth and creamy, effervesce rapidly when poured from the bottle, and have the pleasant, acid taste of buttermilk, with the added sharpness caused by the gas and the trace of alcohol. Kefir 2 or 3 days old may have a yeasty taste, but if it has been properly made this will disappear as the fermentation of the sugar nears completion; made under these conditions, it should be used when 3 to 5 days old, but if put on ice it may be held for a week or even longer.

DETERMINING THE VITALITY OF LACTIC CULTURES

Cultures used for cheese making or cultured milk produce acid at different rates. The following formula is used to determine the vitality of lactic cultures. A pint of milk is placed in a sterile ground stoppered bottle and heated to 100° F. Then 5 cubic centimeters of the lactic culture is added and 30 minutes later 1 cubic centimeter of rennet is added. One hour after adding the rennet the curd is cut and 2 hours after cutting the curd the whey is drained off. At periods of 2 and 3 hours after draining the whey acid tests are made on the whey surrounding the curd. A "slow" culture will produce whey titrating 0.16 and 0.23 while an active culture will produce whey titrating 0.42 and 0.63% of acid at the two consecutive titrations. The titrations are made with tenth normal sodium hydroxide using phenolphthalein as an indicator. The result is expressed as lactic acid.

SOUR CREAM

To 20% cream add 2 to 3% skim milk powder. Heat slowly to 120° F. to dissolve the powder and follow this by pasteurizing at 145° F. for 30 minutes. Cool to 70 to 72° F., and add 3 to 5% of good starter, thoroughly broken up. Dilute 20 drops of commercial rennet extract in about ¼ glass of water and add this to 100 pounds of cream, agitating it thoroughly to distribute the rennet. The rennet helps to form a thick curd and the cream may curdle in a relatively short period. However, you should hold it overnight at the ripening temperature of 70 to 72° F. to develop the desired acid flavor. Follow this by breaking up the curd while cooling to 40° F. and hold at this low storage temperature.

INFANTS' MILK, SYNTHETIC

Sugar	40 g.	Dextrin	20 g.
Soya Bean Powder	125 g.	Egg Yolk, Liquid	50 g.
Lactose	30 g.	Calcium Lactate	6 g.
Peanut Oil	20 g.	Salt	2 g.

Stir in water before use.

SOYA BEAN VEGETABLE MILK

If the dried beans, preferably yellow-seeded varieties, are soaked for a few hours, then finely crushed and boiled for about 30 minutes in the proportion of 3 parts of water to 1 part of mash, a milky emulsion is obtained which is very similar in appearance and properties to animal milk. This liquid, separated out by means of a very fine sieve or cloth strainer, is the Soya Bean or vegetable milk used so extensively in China. Soya bean meal after the oil is extracted or whole soya bean meal may be utilized quite as well as the whole bean. In the absence of animal milk, soya bean milk is used extensively in the fresh state and as the basis of various kinds of vegetable cheeses in oriental countries. Soya bean milk in the form of a powder is a commercial product in some European countries, and in parts of the United States it has been used in special feeding cases. The milk can be used successfully in numerous preparations, such as breads and cakes, in creaming vegetables, in milk chocolate, and in custards.

After separating the liquid from the solid material, the residue is still very rich in nutritive substances and can be dried and used for cattle feed or made into flour for human food.

SOYA BEAN CURD

The addition of magnesium or calcium salts or of rennet or lactic acid to soya bean milk when hot precipitates some of the protein, forming a grayish white curd which settles out, leaving a yellowish water liquid. This curd, after being drained and pressed, represents bean curd of tofu, which is extensively eaten and forms the basis of numerous fermented, smoked, and dried cheeses in China and Japan. Bean curd is made fresh daily and is a staple article of diet among oriental peoples. In many cities of the United States having a large oriental population fresh bean curd may be found in the Chinese and Japanese markets.

DRY MIX FOR MAKING CHOCOLATE MILK IN DAIRIES

Cocoa	1.75 lb.	Vanillin	0.003 lb.
Cane Sugar	7 lb.	Salt	0.025 lb.
Agar, Powdered	0.14 lb.		

Mix the above ingredients well and add to each gallon of milk in the pasteurizer at 185° F. Agitate and hold for ½ hour.

COCOA MALT POWDER

Cocoa Powder	23 lb.	Sodium Bicarbonate	2 oz.
Fine Granulated Sugar	70 lb.	Salt	8 oz.
Malt Powder, Mild Flavor	20 lb.	Vanillin	½ oz.
Skim Milk (Soluble)	14 lb.	Vanilla Extract	½ oz.

Mix ingredients thoroughly and pass through a coarse sieve. This mixture can be packaged in cans, glass containers, or in 1¼-ounce envelopes for individual use.

STABLE CHOCOLATE MILK
U. S. Patent 1,989,758

In carrying out the process of making the milk starch emulsion, the chocolate, sugars (when the latter are used), starch, and the gum may be introduced, as dry substances, into the milk, thoroughly mixed, and the mixture heated to a temperature of 170° to 200° F., or higher if desired—although this is not necessary—in place of temperatures approximating 240° F. heretofore recommended, for periods from 20 to 30 minutes, more or less. Preferably, however, a syrup is first made of the chocolate and sugar, and this syrup, together with a preformed mixture, in proper proportions, of the starch and gum, added to the milk and the final mixture agitated and heated as described.

As a matter of convenience to the beverage manufacturer, and in order to insure correct proportions between starch and gum, the starch and gum may be compounded together and the compound delivered to the beverage manufacturer.

In making the compound the agar-agar, for example, is preferably ground dry and screened to the same degree of fineness as the starch and is then thoroughly mixed with the starch in the proportions indicated by the specific examples given below. In such a mixture the agar-agar, although very small in quantity, approximately from 1 to 20 parts of agar to 100 parts of starch, will remain evenly distributed in the starch. It will not sift out. This novel mixture will disperse in the chocolate vehicle much more easily than if the ingredients were introduced into the liquid as separate substances. If the agar is not finely ground it will swell instead of dissolving, particularly at the low temperatures preferably used in compounding, with consequent loss of stabilizing power.

The following examples of typical mixtures, with preferred percentages of the ingredients, will serve to illustrate the character of the present invention. The percentages are by weight.

Formula No. 1

Milk	90.48
Cane Sugar	4.82
Dextrose (Cerelose)	2.41
Cocoa (High Grade, Dark)	1.27
Raw Tapioca Starch (Scott Test 150)	1
Agar-Agar	0.02

Any suitable sugars may be used in the suspension or in the dry product or the sugar ingredient may be omitted if desired. The amount of the sugar ingredient may be varied to any extent. For any usable quantity the sugar does not add to the viscosity of the beverage. The amount of cocoa or chocolate may also be varied. The matter of taste or of economy will govern any increase or decrease. As much as 2.5% of cocoa may be used without changing the percentage of starch or gum. The starch ingredient may be increased to 2 or 3%. Experience goes to show that 1% is near the critical lower limit. More than 2 or 3% gives too high a viscosity and is likely to give a distinct starch taste to the product. The agar-agar may be varied in amount from about 0.01 to 0.2%, but at the upper limit there is a strong tendency to segregation in jelly-like lumps.

No. 2

Milk	90.78
Cane Sugar	4.06
Cerelose	2.03
Cocoa (Cheaper Quality Than in No. 1)	1.673
Raw Corn Starch (Scott Test 100)	1.433
Gum	0.024

The first four items may be varied as indicated in No. 1.

The same quantity of modified corn starch may be used in place of the specified raw corn starch. The amount of corn starch may vary between 1 and 2%. Where raw corn starch is used the lower limit of the gum quantity should not be quite as low as in No. 1.

No. 3

Milk	91
Cane Sugar	4.07
Cerelose (Corn Sugar)	2.03
Cocoa	1.676
Wheat Starch (Scott Test 85)	1.2
Gum	0.024

The variations may be substantially the same as with No. 1.

The time of cooking with the raw corn starch should be ordinarily 25 to 30 minutes; with the modified corn starch 20 to 25 minutes; with the tapioca and wheat starches about 20 minutes.

CHOCOLATE-FLAVORED MILK

In this improved formula use:

Cocoa	20 lb.
Sugar	90 lb.
Skim Milk	90 lb.

To the above syrup add 2000 pounds of milk; heat to 143½° F. and hold for 30 minutes. Homogenize the mixture at 2000 to 3000 pounds pressure while hot.

Cool and bottle.

NON-SETTLING COCOA MILK

Cocoa Powder	6 oz.
Sugar	28 oz.
Sodium Alginate	1 oz.
Milk	15 qt.

Mix together the cocoa, alginate, and sugar. Heat the milk to 160° F., add the dry mixture slowly with constant stirring, for 30 minutes. Cool the batch to 45° F. and hold for 2 hours before bottling in sterilized bottles. The cocoa powder can be of any fat percentage from 10 to 25%. The milk can be either whole milk or skim milk, or any mixture of each. Additional flavoring ingredients such as vanilla, malt, caramel, etc., may be added.

WHIPPIT

A preparation to aid in whipping cream.

Lime water and sugar, or a saturated solution of calcium saccharate.

Direction: 1 teaspoonful Whippit to the pint of cream.

WHIPPED CREAM IMPROVER

1. Water	7 pt.	Mix sugar and agar, and boil up with water for 3 minutes.
Sugar	3 lb.	
Agar	1 oz.	

2. Water 1 pt. } Cook this a day ahead and add to above. Mix and bring
   Corn starch 4 oz. } to boil.

3. Cream the above cold mix in the beater. Then add 1 lb. meringue powder and beat until fluffy.

The addition of this finished mix, in the proportion of 1 gallon to 3 gallons of whipped cream, results in a product superior to that mixed with straight meringue. The cream should be whipped up first and then incorporated with the above improver.

CREAM TABLES

TABLE FOR MIXING CREAM AND MILK TO OBTAIN 100 POUNDS OF 20% CREAM

When Cream Shows Butter Fat to Be	Take		Take	
	Cream, lb.	Skim Milk, lb.	Cream, lb.	4% Milk, lb.
21%	95	5	94	6
22%	91	9	89	11
23%	87	13	85	15
24%	83	17	80	20
25%	80	20	76	24
26%	77	23	73	27
27%	75	25	70	30
28%	72	28	67	33
29%	69	31	64	36
30%	67	33	62	38
35%	57	43	52	48
40%	50	50	45	55

TABLE TO REDUCE 100 POUNDS OF CREAM TO 20%

To 100 Lbs. Cream Showing	Add		To 100 Lbs. Cream Showing	Add	
	Skim Milk, lb.	4% Milk, lb.		Skim Milk, lb.	4% Milk, lb.
40%	100	125	26%	30	37½
35%	75	93¾	25%	25	31¼
30%	50	62½	24%	20	25
29%	45	56¼	23%	15	18¾
28%	40	50	22%	10	12½
27%	35	43¾	21%	5	6¼

1 gal. 30% Cream weighs 8.3 lb. 1 gal. 40% Cream weighs 8.2 lb.

QUANTITY TO BE TAKEN FROM CREAM AND SKIM MILK TO MAKE 100 POUNDS CREAM TESTING

Cream Shows Per Cent Fat	8%	10%	12%	14%	16%	18%	20%
16	50.0	62.5	75.0	87.5	100.0		
17	47.0	58.8	70.9	82.3	94.0		
18	44.4	55.5	66.6	77.7	88.8	100	
19	42.1	52.1	63.1	73.6	84.2	94.7	
20	40.0	50.0	60.0	70.0	80.0	90.0	
21	38.0	47.5	57.1	66.6	76.1	85.7	95.2
22	36.3	45.4	54.5	63.6	72.7	81.8	91.9
23	34.7	43.4	52.1	60.8	69.5	78.2	86.9
24	33.3	41.6	50.0	58.3	66.6	75.0	83.3
25	32.0	40.0	48.0	56.0	64.0	72.0	80.0
26	30.7	38.4	46.1	53.8	61.5	69.2	76.9
27	29.6	37.0	44.4	51.8	59.2	66.6	74.0
28	28.5	35.7	42.9	50.0	57.1	64.3	71.5
29	27.5	34.3	41.3	48.2	55.1	62.0	68.7
30	26.6	33.3	40.0	46.6	53.3	60.0	66.6
31	25.8	32.3	38.7	45.1	51.6	58.0	64.5
32	25.0	31.2	37.5	43.7	50.0	56.2	62.5
33	24.2	30.3	36.3	42.4	48.4	54.5	60.6
34	23.4	29.4	35.2	41.1	46.8	52.9	58.8
35	22.8	28.5	34.2	40.0	45.7	51.4	57.1
40	20.0	25.0	30.0	35.0	40.0	45.0	50.0
45	17.7	22.2	26.6	31.1	35.5	40.0	44.4
50	16.0	20.0	24.0	28.0	32.0	36.0	40.0

The figures 8, 10, 12, 14, 16, 18, 20 represent the per cent of butter fat desired.

The other figures represent the pounds of cream to be taken, and this number subtracted from 100 gives the pounds of skim milk required.

EXAMPLE Cream on hand shows 26% Fat.
Cream to be made shall show 20% Fat.

Find 26% in the left hand column and follow the horizontal line to 20% and you come to the number 76.9.

You must take 76.9 pounds of your 26% and add to it 23.1 pounds of Skim Milk to make 100 pounds 20% Cream

$$\begin{array}{r} 100 \\ -\ 76.9 \\ \hline 23.1 \end{array}$$

COMPOSITION OF MILK, MILK PRODUCTS AND BUTTER

Kind	Water %	Butter-fat: %	Protein: (Casein and Albumin) %	Milk Sugar: %	Minerals: %	Total Milk Solids: %
Liquid Whole Milk	88	3½	3¼	4½	¾	12
Evaporated Whole Milk	72	8	7¾	10½	1¾	28
Sweetened Condensed Milk (Cane Sugar)	31	8 41	7¾	10½	1¾	28
Powdered Whole Milk	1½	27½	27	38	6	98½
Liquid Skimmed Milk	91		3½	4¾	¾	9
Skimmed Sweetened Condensed Milk (Cane Sugar)	29	43	11	14½	2½	28
Powdered Skimmed Milk	2½	1½	38	50	8	97½
Liquid 18% Cream	75	18	2½	4	½	25
Powdered Cream	½	72	10½	14½	2½	99½
Butter from Fresh Cream	13.03	83.75	0.64	0.35	0.14 2.09 Salt	
Butter from Ripened Cream	13.78	82.97	0.84	0.39	0.16 1.86 Salt	

CHEESE

A food product obtained by coagulating the casein of milk and incorporating a large proportion of the fat. The usual way of making cheese is to allow "rennet" (a soluble ferment found in the fourth stomach of calves) to act upon milk. One part of powdered rennet will often coagulate 250,000 parts of milk. In order that cheese of a regular quality is obtained, a starter is generally used to ripen the milk before renneting. This starter is a pure culture of lactic acid. To ascertain whether the milk is ripe for renneting a Rennet-Test is made, which consists of taking 4 ounces or 140 cubic centimeters of the milk and adding ½ ounce or 5 cubic centimeters of a diluted rennet extract. The mixture is stirred and the time taken for coagulation is noted. For cheddar cheese the time required is from 45–60 seconds. When the milk coagulates more slowly it is not sufficiently ripe for renneting.

The striking difference in the flavors and consistencies of cheese are traceable to variation of one or several of the following factors:

1. Type of milk used.
   a. Whole milk
   b. Skimmed milk (or partially skimmed)
   c. Cream
   d. Combination of milks

2. Source of milk product.
   a. Cows
   b. Goats
   c. Sheep
   d. Combination of milks
3. Strain of bacteria, mold, etc., used in the ripening process.
4. Amount of ripening and aging to which the product is subjected.

American Cheese Also known as American Cheddar or Store Cheese	Made from sweet milk, whole or skimmed, in large molds, weighing from 10 to 60 pounds, and of varying shades of yellow or uncolored. Originally made in Cheddar, England, but now imitated in the U. S. A. Many varieties of packaged cheese are made from it.
Brick Bat	An English cheese made from new milk and pressed into a brick-like form.
Brie Cheese	A richly salted cream cheese, made in Brie, France
Camembert Cheese	A finely flavored cream cheese of a pale yellow color, made in France. Usually put up in small round wooden boxes and in round tins.
Cottage Cheese, Clabber Cheese, Dutch Cheese, Pot Cheese, Smeerkaas	A pressed curd, into which cream is added, without rennet. Cottage cheese of the best texture; contains 70 to 75% of water.
Cream Cheese	A soft cheese made from unskimmed milk, with or without the addition of cream. The name does not imply that the cheese is made from cream.
Dunlop Cheese	A soft cheese similar to single Gloucester, made in Ayrshire, Scotland.
Dutch Cheese	a. A hard cheese made from skimmed milk pressed into round molds, the outside being colored a reddish tint with a preparation of madder. b. The same as cottage cheese.

Edam Cheese	A hard mild cheese made in Holland Molded in round forms and usually coated with a dark red color or with paraffin. Each cheese weighs from 3 to 4 pounds. Packed 12 pieces to the case.
Filled Cheese	A kind of cheese made from skim milk, so called because a fatty filling such as lard or oleomargarine is used as a substitute for the cream.
Gloucester Cheese	A rich English cheese of mild flavor. It is called single or double Gloucester, depending on the amount of cream in its composition.
Gorgonzola Cheese	Is a rennet cheese made from fresh whole cow's milk in Northern Italy. It takes its name from the village of Gorgonzola, a few miles from Milan, but the manufacture of the cheese has spread over a wide area. At its best, Gorgonzola is very similar to Roquefort.
Gruyère Cheese Schweitzer, Swiss or Emmenthaler Cheese	A cheese made in France and Switzerland, intermediate between hard and soft, yellow, and of an open structure. The characteristic holes of this cheese are caused by gas-producing bacteria during the process of ripening. It is a mild cheese, dry, and of a distinctive, somewhat sweet flavor. A "Domestic Swiss" is made in the U. S. A. but is inferior to the imported cheese.
Limburger Cheese	A soft cheese originally made at Limburg, Belgium, and now in the U. S. A. It is kept until putrefaction begins, hence its strong odor and taste. Put up in 1- and 2-pound bricks. Packed in full cases, weighing about 125 pounds, half and quarter cases.
Lincolnshire Cheese	An English cheese made in small molds from new milk and cream.
Neufchâtel Cheese	A delicate cream cheese made in small molds at Neufchâtel, France.

Parmesan Cheese	A hard dry cheese of delicate flavor and colored with saffron. Made in Italy. When broken, it has a granular appearance. Used grated for seasoning macaroni. Long keeping qualities.
Pont l'Evêque Cheese	A cream cheese similar to Neufchâtel cheese.
Primost Cheese	A cheese made from whey and brown sugar, a by-product of American cheese. Put up in 2½-pound pieces. Especially liked by the Scandinavian people,
Roquefort Cheese	A hard cheese made at Roquefort, France, from ewes' milk. When the cheese is almost solid and dry it is placed in caves in which the temperature remains at 40° F. Here the cheese takes on a reddish color and develops the green mold around the bread crumbs used in preparing it.
Sage or Green Cheese	Regular American cheese to which have been added leaves of the sage to impart a green color and give it the sage flavor.
Sap Sago Cheese	A hard skim milk cheese, made in conical molds, weighing about 6 ounces. It contains a powdered herb of the clover type, which gives it its characteristic flavor and green color. The sap sago cheese is used for grating purposes mainly and is especially liked by the German people (Kraeuter-Kaese).
Slipcoat Cheese	A rich cheese made from milk while it still retains its natural warmth. It is soft and resembles uncolored butter.
Stilton Cheese	A double cream cheese of strong taste. Originally made at Stilton, England.

The Government grades of most general interest to the consumer are:

1. U. S. Extra Fancy	Score 95 or above
2. U. S. Fancy	Score 92–94
3. U. S. No. 1	

Maximum Score

Body and texture	40
Flavor	30
Finish and appearance	20
Color	10
Total	100

VARIATIONS IN COMPOSITION, AGING, ETC.

Kind	Aging	Approx. Content of			Milk Product Used
		Water	Fat	Protein, Ash	
American Cheddar	Unripened to fully ripened	36%	34%	30%	Whole cow's milk.
American Cheddar	Unripened to fully ripened	57%	5%	38%	Skimmed cow's milk.
Swiss	Ripened	33%	30%	37%	Whole cow's milk.
Cream	Unripened	39%	47%	14%	Cream or enriched cow's milk.
Cottage	Unripened	72%	1%	27%	Sour skimmed cow's milk.
Roquefort	Ripened	38%	32%	30%	Cow's milk.

RED CHEESE RIND COLOR

Formula No. 1

Sudan 4 dye is dissolved in equal parts of 70% alcohol and acetone, or

No. 2

Tournesol, Fuchsin, or Bordeaux Red dissolved in water (distilled water is preferred), or

No. 3

Iron Oxide, known also as Berlin Red or English Red, made into a paste with a heavy oil.

The intensity of the color can be varied by changing the amount of the coloring substance.

Apply to outside of cheese.

CHEESE, ICE-CREAM, AND SALAD STABILIZER

U. S. Patent 2,007,218

Locust Bean Gum	65 oz.
Irish Moss, Powdered	35 oz.
Karaya Gum	15 oz.

When used in the preparation of cream cheese, the undiluted mixture of the three ingredients mentioned above is added at the time that the curds are mixed with the cream in the usual procedure for the manufacture of cream cheese, and in the proportion of about one-half of 1% by weight on a wet basis. The material is heated to about 165° F., homogenized, and then packed hot.

In ice cream it is used diluted with sugar, in the preferred proportion of one-half of 1% on a wet basis. The stabilizer acts to prevent crystallization of ice particles and thus insures a fine, smooth texture and a body which will hold up under severe shocks, such as are encountered in transportation and handling. The use of it in ice-cream also usually results in more rapid freezing, especially in old-style freezers.

CHEESE EMULSIFIERS
U. S. Patent 1,940,031

1–4% of either of following are used: Sodium Mucate, Sodium Lactate.

PRESERVATION OF RINDLESS CHEESE
British Patent 434,371

Bacterial action on surface of rindless cheese is prevented by treatment with following prior to heating to 65° F.:

Hydrogen Peroxide (35%)	0.3%

Low fat content cheese is heated to 65° C. The peroxide is added, mixed, and later heated to 80° C.

BRANDY CHEESE

Use regular cheddar cheese, preferably an entire small cheese with the surfaces scraped clean, and allow to dry at room temperature for 2 to 4 weeks. Then place cheese in clear water at 40 to 80° F. for several days.

The cheeses are then placed in a mixture of brandy and high-grade vinegar for several days. The brandy may be mixed in equal parts or less with the vinegar. Three per cent of salt should be added with a liberal addition of pepper to the brandy-vinegar solution.

CHAPTER II

CACAO, CHOCOLATE PRODUCTS,* COFFEE AND TEA

Cacao is derived from the seeds of Theobroma CACAO. The cacao (often incorrectly called cocoa) tree is indigenous to Brazil, Venezuela, Ecuador, and Mexico, thriving only in tropical climate in localities that have well-watered, well-drained soils, and plenty of rainfall. It is a medium-sized tree, 20 to 30 feet high. The fruit is a pod, having a hard, thick shell, and is from 8 to 10 inches long and from 3 to 4 inches thick. There are some 40 or more beans in each pod.

DIFFERENT KINDS OF CACAO BEANS

Porto Cabello or Caracas

Large, but not very uniform in size.

Covered with a reddish, sometimes silver-gray dust.

The husks are thicker than on ordinary beans and are of dark red color inside. Ninety-nine hundredths of one per cent theobromine. The small Caracas bean has less fat and is therefore cheaper.

Trinidads

The beans are nearly flat, the husk is dark gray and the kernels almost black. Ninety-two hundredths of one per cent theobromine. A good bean for the price.

Guayaquils (Ecuador)

Large, flat, irregular shaped beans. Husk and kernel are brown. Fine strong flavor.

Cubans

Egg-shaped, flat beans. Husk mauve, kernel dark brown.

Paras (Brazil)

Irregular in size, long and flat on one side. Husk gray or reddish, the kernel light brown. Rich in fat, good for filler.

Grenadas

Similar to Paras.

* Courtesy of B. J. Zenlea, Baker Chocolate Co., and Food Industries, N. Y. City.

Javas

Oval, slightly flattened beans. The husk thin and easily peeled. The kernel is light brown, of bitter but agreeable taste. One and twenty-two hundredths per cent theobromine. Fine flavor, but high in price.

Bahias

Flat, triangular in shape, and rough. The yellow looking bean is the best, the grayish-red one inferior. The kernel is mauve in color and produces a dark liquor. A very popular bean in the United States.

TEST FOR RIPENESS OF A CACAO BEAN

Cut the bean with a knife. If it slices and no shell falls off, the bean is unripe.

THE (COCOA) CACAO BEAN AND ITS PRODUCTS

Cocoa has been cultivated since very ancient times, but this plant and its products were unknown to Europe until the beginning of the sixteenth century, when Cortez and his soldiers became acquainted with it upon their landing in Mexico. From then on the use of cocoa spread very rapidly throughout the civilized world, until today cocoa is of international usage.

Cocoa is grown almost entirely south of the Equator, the largest crops coming from the west (Gold Coast) of Africa. The next largest crop comes from Ecuador.

The cocoa from each country has a distinctive character and flavor of its own. The cocoas coming from Venezuela are the finest and the cheapest are from San Domingo.

Cocoa beans are the seeds of a tree that grows in most tropical countries. The seeds grow in pods very much like the seeds of the pumpkin and squash, and hang from both the branches and trunks of the trees. When ripe they are picked from the tree and gathered into large piles. They are then cut open and the seeds and pulp scraped out.

The seeds with their surrounding pulpy mass are placed into large vats or pits and allowed to ferment. The time of fermentation varies according to the kind of cocoa. The sweeter variety requires a shorter fermentation while the coarser or more ordinary type requires a longer fermentation.

During the fermentation starch is converted by the diastase present into soluble sugars, which are further fermented into alcohol, and finally converted to acetic acid. The pulpy mass is decomposed and flows away as a vinegar-smelling fluid. During the fermentation care must be taken that sufficient air is circulated through the beans, otherwise, if insufficient oxygen is present butyl alcohols are formed which will give the cocoa a bad odor. To prevent this the beans are turned over every day. Fermentation changes the color of the bean from a light to a dark rich red brown, liberates the alkaloid theobromine, and lessens its bitter and astringent taste.

When the fermentation is completed the beans are separated from any remaining pulp, dried, and sometimes polished. The drying must be done very carefully, as improperly dried beans are apt to become moldy. The best cocoa is obtained by drying in the sun, although methods using steam- or water-heated pipes are also in use.

Upon importation into the United States, the beans are inspected and are rejected if they contain more than 5% of either wormy or moldy beans.

The raw cocoa beans are cleaned by passing them over a shaking-table covered with a metal screen which has holes smaller in size than the beans. Sand, dust, and other foreign materials are screened out. At the end of the shaking-table, the beans drop into a chute located at the bottom of an air duct. The cleaned beans are carried up by air suction, and deposited into a conveyor. Stones, metals, and other heavy materials drop to the bottom of the chute, which is counterpoised by a weighted lever. When sufficient heavy material has gathered, the bottom opens and the foreign material drops to a receiving pan.

The clean raw cacao beans are run into a roaster of approximately 600 pounds capacity, of the revolving drum type, which is heated by coal fire using forced draft, or by gas or superheated steam. A roast usually requires 45 minutes, depending on the type of cacao beans, and the kind of roast desired. The final flavor of the cocoa powder and chocolate depends a great deal on the proper roasting of the cacao beans.

The cacao beans attain a temperature from 240° to 300° F. at the end of the roast, are quickly discharged into a cooling truck with a perforated metal bottom, and are quickly cooled by sucking cold air through them.

The cooled beans are carried to a cracker and fanner device, which cracks up the beans and blows the shells aside. The cocoa germs and fines are separated from the broken nibs by sieving them through a rotating cylinder covered with screening of increasing mesh. The clean nibs are carried to hoppers which feed a battery of triple stone mills, and are ground into a heavy, viscous fluid known as chocolate liquor, and is the starting point in the manufacture of cocoa powder and chocolate.

In the manufacture of cocoa powder the chocolate liquor is transferred to hydraulic presses, and pressed from 8 minutes to 1¾ hours, depending on the amount of fat that is to be left in the cocoa powder. The cocoa press cake is cooled, and broken up by a cracking device, passed over a magnetic belt to remove stray iron, and then put through a micro-pulverizer. The ground press cake is cooled on a conveyor, and refined by sieving through silk bolting cloth, or by playing a current of air on the particles of cocoa in such manner as to separate the fine cocoa from the coarse. The coarse cocoa, or cocoa tailings, are returned to the grinders and reground over and over again, until the particles are fine enough to pass through the sieve, or can be separated by air flotation.

There are many kinds of cocoas, but they can be divided into two main divisions, namely,

a. Natural process cocoa

b. Dutch process cocoa

Natural process cocoas are cocoas to which no alkalies or "Dutching Agents" have been added during manufacture.

Dutch process cocoas are cocoas to which an alkali has been added. In preparing Dutched cocoas, the clean nibs are boiled in a solution of alkali, such as potassium, sodium, ammonium bicarbonate, or carbonate. The water is evaporated and the Dutched nibs dried in a roaster. The nibs are milled, the liquor pressed, and the cocoa refined in the same manner as for natural process cocoas.

Dutched cocoas are darker in color, less acid, and less bitter than natural cocoas. They are more economical to use, and, for certain purposes, cannot be replaced by natural cocoas.

The various types of cocoas can be described as follows:

1. Semi-cocoas, cocoa nuggets, cocoa crumbles, cocoa chips, etc., are prepared by crumbling press cake or by mixing chocolate liquor with a quantity of cocoa to reduce the fat. In the latter case, the product is molded into thin cakes, and then broken into small chips. Because of the high fat content, 32 to 45%, and chocolaty flavor, they are used extensively for fine quality ice-cream, and for fancy icings.
2. High-quality cocoas usually contain from 25 to 32% cocoa butter. They are made from select quality cocoa beans of the better type, and are carefully blended. These cocoas are either Dutched, natural, or semi-Dutched. They are used for fancy ice-cream, for drinking, for icing, as a filler between layers of cakes, and for syrups.
3. Breakfast cocoas contain not less than 22% cocoa butter. The natural process and semi-Dutched breakfast cocoas are used for drinking, syrups, and general household use. The Dutched type breakfast cocoas are used for chocolate biscuits and crackers, for devil food cakes, and for home candy making, such as chocolate fudge.

4. Semidry cocoas usually contain 15 to 18% fat. The natural and semi-Dutched cocoas are used for malted milk mixtures, cocoa milk drinks, chocolate flavored desserts, chocolate pies, and chocolate flavored syrups. The percentage of fat they contain is just right for use in the products mentioned.
5. Dry cocoas contain 10 to 12% fat, and are used for miscellaneous purposes. The natural process cocoa is used for milk drinks, cocoa skim milk mixtures, sweet cocoas, icings, etc. The Dutch process cocoa is used for cocoa extracts, for baking, and devil food mixtures. These cocoas are the most common type found on the market, and are used where low cost is of importance.
6. Extra dry cocoas contain between 8 and 9% cocoa fat. They are used for thin syrups, and other mixtures where any excess fat is not desired.
7. Very dry cocoas contain 6½ to 7¾% fat, and this fat content is about the limit for cocoas made by pressing the fat from the chocolate liquor. They are used for making substitute or imitation chocolate coatings by incorporating sugar and another fat, such as cocoanut oil. They are desired for their low fat content, because of the difficulty of blending cacao butter and cocoanut oil. This cocoa is used as a dusting powder in confectionery.
8. Solvent-extracted cocoas or defatted cocoas contain from ½ to 3% fat. They are usually manufactured from cheap low-fat cocoa cake, by extracting the remaining fat by means of a solvent. Solvent-extracted cocoas have practically no chocolate flavor. Some have a faint odor and taste of the solvent used for their extraction. They are usually high in moisture content, running at times as high as 10%. Solvent-extracted cocoas are used as dusting powders in confectionery, and for imitation chocolate.
9. By-product cocoa and cocoa cake usually contain 5½ to 11% cocoa fat. This product is obtained by passing cocoa fines and dust through the Anderson Expeller, which removes a portion of the fat. The by-product cake has a limited use as a fertilizer ingredient, as raw material for theobromine extraction, and for cattle feed. Its use for the latter purpose is limited, and has not been very successful.
10. Prepared cocoas are cocoa mixtures, such as cocoa and sugar, cocoa skim milk and sugar, and cocoa malt skim milk and sugar. They are used for instant chocolate and milk drinks in ice cream parlors, and are also put up in envelopes of the proper weight sufficient for one cup.

THE MANUFACTURE OF CHOCOLATE

Chocolate, because of its irresistible flavor, its concentrated food value, pleasing appearance, keeping qualities, and convenience in handling, is prized by the young, the old, the athlete, the warrior, and the explorer.

The definition of chocolate, in a technical and restricted sense, is the solid or plastic mass obtained by grinding cacao nibs free from shell or husk and germs, and is the product commonly known as chocolate liquor. By common usage, however, the word chocolate is usually applied to sweet chocolate, milk chocolate, and chocolate covered centers.

Although chocolates may be similar in composition as to percentages of chocolate liquor, sugar, added cocoa butter, and milk solids, there is a world of difference in their flavor and texture.

There are many varieties of cocoa beans, of which approximately thirty are used in the larger chocolate factories. The cocoa beans vary as to color, chocolate character, pungency, acidity, mildness, bitterness, and creaminess. The delicate flavors of the cocoa beans are developed by careful roasting, cleaning, and milling. The flavor of the resulting chocolate is dependent upon the skill used in blending, milling, and conching the various varieties of cocoa nibs with the proper amount of sugar and cocoa butter, with or without milk solids, and flavors.

The possibility of developing blends of chocolate is unlimited. Most formulae are limited to blends of one to six varieties. The number of combinations that can be made with thirty varieties of cocoa beans blended with varying percentages of sugar, cocoa butter, milk solids, and flavors, and milled to various degrees of fineness, and adjusted to various viscosities and colors, is astounding.

There are many varieties of chocolates on the market, but they can be classified into the following general types: Bitter, bittersweet, dark sweet, medium dark sweet, light sweet, milk chocolate, ice-cream coating, skim milk, buttermilk, malted milk, imitation, and lecithin coatings.

It is difficult to state exactly what type of chocolate coating is best suited for a particular center. It depends on a number of factors among which are sweetness, texture, and blending properties of both the center and coating. As a rule chocolate liquors and bittersweets are used for every sweet center. Dark sweets and medium dark sweet coatings are used for moderately sweet centers. Milk chocolates are used for caramels, nougatines, marshmallows, etc. Light sweets are used for centers not too sweet and for candy bars.

The chocolate liquor of commerce is also known as chocolate, bitter chocolate, chocolate liquor, chocolate paste, bitter chocolate coating, and unsweetened chocolate. It is the product obtained by grinding cacao nibs free from shell or husk and germs, usually through triple stone mills, Bausman disc, or similar devices.

The above product comes in two types, i.e., natural process and Dutch process. The Dutch process type, or alkalized type, is prepared in the same manner as described for cocoas except that none of the fat is expressed. The natural chocolate liquor has a *p*H value of 5.1 to 6.0. The Dutched chocolate liquor has a *p*H value of 6.0 to 7.8, depending on the original *p*H of the nibs, and the amount and kind of alkali used. If more than 1½% alkali is used the liquor has a tendency to develop a saponified taste.

Chocolate liquors are molded into ten-pound cakes for commercial use, and in one-quarter to one-pound cakes for household use. Chocolate liquor is used for coating very sweet centers, such as hand roll centers and butter creams. It is also used as an ingredient for ice cream, candy, culinary products, and chocolate syrup. The Dutched type of chocolate liquor is used for flavoring tobacco, and for general uses where a milder and a darker product is desired. Chocolate liquor is often used in place of cocoa powder when a creamier and richer product is desired. The high fat content, 50 to 56%, however, makes its use difficult without such aids as homogenizers or emulsifiers.

Chocolate liquor is considered pure if it contains not less than 50% cocoa fat, and on the moisture and fat-free basis not more than 8% of total ash, not more than 0.4% of ash insoluble in hydrochloric acid, and not more than 7% crude fiber. The standards for Dutched liquor are the same as for natural process liquor, after due allowance being made for the kind and amount of alkaline substance added. The amount of alkaline substance permitted is not more than three pounds of potassium carbonate, or its neutralizing equivalent in other alkaline substances to each 100 pounds of cacao nibs.

Sweet chocolate, sweet chocolate coating, is chocolate mixed with sugar and/or dextrose with or without the addition of cacao butter, spices, or other flavoring materials. This definition applies to all sweet coatings regardless of the amount of sugar and chocolate liquor they contain. The phrase, other flavoring ingredients, is somewhat vague, and has led to some misunderstanding. The definitions for food products are so framed as to exclude anything not mentioned by name, but the strict interpretation of "flavor" would include any and all flavoring ingredients such as small amounts of malt, malted milk, skim milk, dairy butter, essential oil, etc.

Although dextrose (corn sugar) is permitted in chocolate, either alone or mixed with sucrose (cane sugar or beet sugar), it has been used only to a very limited extent, and has not been studied sufficiently to denounce or extol its virtues.

Bittersweet coatings contain from 80 to 95% chocolate liquor and 5 to 20% sugar. Since the cacao butter content is very high, 40 to 50%, and the viscosity low, the addition of extra cocoa butter is not required. These coatings are prepared by adding sugar to chocolate liquor with or without flavor and refining. Bittersweet chocolates are dark in color and are used as couvertures for cordials, peppermint creams, fudge, whipped cream, Italian creams, hand rolls, butter creams, and marshmallow centers.

Mild bittersweet coatings contain from 50 to 75% chocolate liquor, from 25 to 38% sugar, and from 38 to 42% cacao butter. As the sugar content increases and the chocolate liquor content decreases, it becomes necessary to add additional cacao butter to obtain the proper viscosity. Mild bittersweet coatings are used for practically the same type of centers as bittersweet coatings, and are used to some extent for covering marshmallow crackers.

Dark sweet chocolates contain from 25 to 45% chocolate liquor, from 40 to 50% sugar, and vary in total cocoa butter content from 34 to 42%, depending on the desired viscosity of the finished coating. Dark sweet chocolates are used for enrobing such centers as nougats, cordials, peppermints, fudge, and nut-flavored cream fondants.

Medium dark sweet coatings are the most popular, being used to a great extent for molding chocolate bars and a variety of solid chocolate items for the retail trade, which are neither too sweet nor bitter for the average consumer's taste. They contain from 20 to 40% chocolate liquor, from 45 to 55% sugar, and from 32 to 42% total cocoa butter. The commercial product is sold in molded ten-pound cakes, and is used for enrobing such centers as caramels, nougats, fruit-flavored cream fondants, jelly centers, cocoanut cream fondants, vanilla cream fondants, nut clusters, raisin clusters, dates, figs, marshmallows, nutflavored cream fondants, cocoanut, sponge, chip and marshmallow cracker.

Light sweet coatings are of two types, the expensive type and the cheaper type. The expensive types are prepared from blends of light-colored and mild but chocolate-like cacao nibs such as Java, Ceylon, Maracaibo, etc., and a moderate amount of sugar, usually 45%. These coatings are very expensive, and are used for enrobing moderately sweet centers of the better sort. They are finely milled, conched, delicately flavored, and are representative of the chocolate expert's skill in the art of fine chocolate making.

The cheap type of light sweet chocolate coatings owe their light color to the small amount of chocolate liquor and the large amount of sugar used in their manufacture. They usually contain from 5 to 15% chocolate liquor, from 55 to 68% of sugar, and contain a total cocoa butter content of 31 to 39%, depending on the fineness of grinding and their desired use. These coatings are used for special purposes, and for enrobing nut rolls, nuts, nut clusters, nut chews, raisin clusters, and for candy rolls containing peanuts, marshmallows, caramels, fudge, and other ingredients. The light sweet coating, though lacking in chocolate flavor and being excessively sweet, nevertheless, blends in nicely with the types of centers mentioned.

Milk chocolate, sweet milk chocolate, is the product obtained by grinding chocolate with sugar and/or dextrose, with the solids of whole milk, or the constituents of milk solids in proportion normal for whole milk, and with or without cacao butter, and/or flavoring material. It contains not less than 12% of milk solids. Milk chocolates usually contain 7 to 17% of chocolate liquor, 35 to 55% sugar, and a total fat content of 28 to 39%. There are many types of milk chocolate, varying in their degrees of milkiness, chocolate character, color, fineness, and viscosity. Milk chocolates are prepared from fresh whole milk which is first evaporated, and from milk powders, both spray and roller process types, and from skim milk powder to which has been added an amount of milk fat either in the form of dairy butter, or butter fat, necessary to convert the skim milk solids to whole milk solids.

In converting skim milk solids to whole milk solids by the addition of dairy butter, it is necessary to bear in mind that dairy butter contains 80 to 84% of milk fat, and approximately 16% water. The difficulties of removing the water and blending of the dairy butter with the skim milk solids, and the fact that 100 pounds of dairy butter yields approximately 80 pounds of milk fat, makes this method of milk chocolate preparation undesirable, and often more expensive than with the use of whole milk solids.

Milk chocolates owe their popularity to their delicate milk flavor, and lack of harsh or bitter taste. They are particularly liked by women and children. The manufacture of milk chocolate with an irresistible milky flavor is a difficult art, and is a secret closely guarded by chocolate manufacturers. Milk chocolates are sold over the counter of the ice-cream parlors and candy stores as broken milk chocolate, and in the form of bars which vary from a fraction of an ounce to a pound. Milk chocolate coatings are molded in ten-pound cakes and are used for enrobing such centers as nougats, vanilla cream fondants, nut rolls, nuts, nut clusters, and nut chews.

Ice-cream coatings are chocolate coatings of high cacao butter content containing from 50 to 60% fat. Their high fat content is necessary to obtain a very thin running chocolate which will quickly drain off the solid ice-cream center when dipped. Ice-cream coatings may be of the dark sweet, medium dark sweet, light sweet, and milk chocolate types, due allowance being made for the high fat content. As their name implies, they are used exclusively for dipping ice-cream in the manufacture of ice-cream pops and eskimo pies. Since ice-cream coatings are frozen instantly on the ice-cream, and the same is kept at temperatures below freezing, the user of ice-cream coatings is not concerned with exact enrobing temperatures, nor with fat or sugar bloom. His chief problem is to combat the thickening up of the chocolate due to moisture absorption from the melting ice-cream. This difficulty is often remedied by the incorporation of three-tenths of one per cent lecithin in cacao butter.

Skim milk, buttermilk, milk chocolate coatings are sweet chocolate coatings containing not less than 12% of, respectively, skim milk solids, buttermilk solids, or malted milk and milk solids. These coatings are not defined in F.I.D. 191, or later additions to the Federal Pure Food Laws. Buttermilk coatings, and skim milk coatings, have become popular in the last few years. They are a cross between milk chocolate coatings and light sweet coatings. Skim milk and buttermilk cut down the excessive sweetness of a light sweet coating without sacrificing the light color. These coatings are used for enrobing the same types of centers as light sweet coatings, and in some cases milk chocolate coatings.

These coatings, in absence of a published definition for the same, should be labeled and sold for exactly what they are, namely, skim milk, buttermilk, or malted milk chocolate coatings.

Modified chocolate coatings are sweet chocolate coatings of various types, such as dark sweets, light sweets, and ice-cream coatings, to which has been added a small amount of skim milk, buttermilk, whole milk, or malt sugar. The amount added is usually 3 to 5%. These ingredients are added as flavoring ingredients, and to cut down the excessive sweetness of a light sweet coating, and also to cut the bitterness or sharpness of a dark sweet coating. The above ingredients are used to some extent in ice-cream coatings for the reasons mentioned, and for obtaining a softer eating chocolate of less brittleness than the ordinary chocolate coating, when same is frozen on the ice-cream.

Imitation chocolate coatings are mixtures of cocoa, sugar, and a substitute fat such as cocoanut oil, cocoanut stearines, illipe oil, etc., with or without flavoring ingredients. They are used mostly for enrobing cheap centers, and for medicated products where a higher melting point chocolate is desired than is obtained with cacao butter. Cocoanut oil coatings are used to some extent for ice-cream. Imitation coatings are what their names imply, imitation chocolate coatings often lacking the true flavor and taste of pure chocolate.

Lecithin chocolate coatings are chocolate coatings of all types to which has been added a small amount of lecithin, usually not exceeding four-tenths of one per cent of lecithin in cocoa butter.

Lecithin modifies the physical properties of chocolate to some extent. Coatings can be made with three to eight pounds less of added cocoa butter per 100 pounds of chocolate, making them less fatty. Lecithin coatings resist thickening up due to moisture, allow greater flexibility in enrobing temperature, and cover centers more uniformly. Lecithin stabilizes chocolate coatings so that the viscosity is only slightly changed on aging, heating, or mechanical manipulation.

Cacao butter, cocoa butter, is the only fat common to chocolate, and is the edible fat obtained from sound cacao beans either before or after roasting. It is a yellowish-white solid having a faint, agreeable odor, and a bland chocolate-like taste, and is usually brittle at temperatures below 25° C. (77° F.). It is slightly soluble in alcohol, but is soluble in ether, chloroform, petroleum benzin, benzene, and in many other solvents. It has a specific gravity of approximately 0.8580 at 100° C. (212° F.), and 0.9100 at 40° C. (104° F.), when compared with an equal volume of water at 25° C. (77° F.). Cacao butter melts between 30° C. and 35° C. (86° F. and 95° F.), has a refractive index of approximately 1.4578 at 40° C. (104° F.), a saponification value of 188 to 195, an iodine value of 33 to 38, and a free fatty acid content of 0.70% to 1.4% calculated as oleic acid.

Cocoa butter is obtained by pressing chocolate liquor in the manufacture of cocoa powder, by passing cacao fines through the Expeller, and by solvent extractions from low fat cocoa press cake and cacao fines.

Cacao butter is used mostly for the manufacture of chocolate and confectionery, and to some extent for the manufacture of suppositories, creams, lotions, ointments, and soaps, and is also used for the treatment of burns, sunburn, and for massaging.

Chocolate, when properly molded, possesses a good gloss which, if not protected, soon becomes dull in appearance. Both heat and moisture rapidly destroy the luster of chocolate, and develop a mottled and gray surface commonly known as bloom.

Chocolate must be carefully melted, carefully tempered, and carefully cooled when manufactured, otherwise bloom may develop in a short time. The handling of chocolate is a fine art and requires very careful attention to details. If chocolate is cooled too rapidly, the outer surface will solidify, trapping the latent heat within, which heat will gradually find its way to the surface and cause fat bloom. The proper temperature for storing chocolate is 55 to 65° F., and the proper relative humidity 60 to 65%.

If chocolate is taken out from a cold room into a warm room, or placed in a damp room, a film of moisture will form on the surface. The fine sugar particles will dissolve and form a thin layer of syrup. On evaporation of the moisture, the sugar crystallizes in minute form, giving the chocolate a rough, grayish appearance, sometimes mottled by red-brown colored spots due to the coloring matter of the chocolate that may have dissolved in the moisture. This phenomena is known as sugar bloom.

If chocolate is stored in a warm place or exposed to heat, the cacao butter softens or melts, and expands, forcing its way through the spaces between the sugar and cocoa matter particles to the surface. On cooling the cacao butter crystallizes, forming a grayish-white film which is greasy to the touch. This type of bloom is known as fat bloom, and is a source of trouble during the summer months. Although the flavor is usually not impaired, the grayish appearance of the chocolate is considered by the public to be a sign of staleness, and results in great economic losses to the manufacturer. Chocolate picks up foreign odors readily, and should be stored in a clean room with free circulation of air.

Cocoa shells are a by-product of the cocoa and chocolate industries. They are used chiefly as a fertilizer ingredient, and also for the extraction of its fat and theobromine content.

Uses	Semi-Cocoas, Cocoa Nuggets, Cocoa Crumblers, Cocoa Chips, 32–45% Fat		High Quality Cocoas, 25–32% Fat		Breakfast Cocoas Not Less than 22% Fat		Semi-Dry Cocoas, 15–18% Fat		Dry Cocoas, 10–12% Fat		Extra Dry Cocoas, 8–9% Fat	Very Dry Cocoas, 6½–7¾% Fat	Defatted or Solvent Extracted Cocoas, ½–3% Fat	By-Product Cocoa and Cocoa Cake, 5½–11% Fat	Prepared Cocoas Sweet Cocoas
	Natural	Dutched	Natural	Dutched	Natural	Dutched	Natural	Dutched	Natural	Dutched					
Baking					●	●	●	●	●	●					
Beverages			●	●	●	●	●	●	●	●					
Cake Fillers			●	●											
Cattle Feed														●	
Chocolate Biscuits						●		●							
Chocolate Crackers						●		●		●					
Choc. Flavored Desserts					●	●	●	●	●	●					
Chocolate Pies					●	●	●	●	●	●					
Cocoa, Malt, Sugar Mixt.							●	●	●	●					
Cocoa Milks							●	●	●	●	●	●			
Confectionery, Fudge			●	●	●	●	●	●	●	●					
Devil's Food Cake						●		●		●					
Dusting Powder												●	●	●	
Fertilizers														●	
Household Use					●	●			●	●					
Ice-Cream	●	●	●	●											
Icings	●	●	●	●											
Imitation Choc. Coatings									●		●	●			
Medicinal Powders or Tablets					●	●	●	●	●						
Sweet Cocoas					●	●	●	●	●	●					●
Syrups			●	●	●	●	●	●	●	●	●				
Theobromine													●	●	
Thin Syrups									●		●	●			
Tobacco			●	●	●	●	●	●	●	●					

USES FOR CHOCOLATE

	Chocolate Liquor	Bittersweet	Dark Sweet	Medium Dark Sweet	Light Sweet	Milk Chocolate	Ice Cream Coating	Skim Milk, Buttermilk	Imitation Coating	Cacao Butter
Butter Creams	●	●	●							
Candy Bars				●	●	●		●		
Caramels				●	●			●		
Chocolate Bars				●	●	●				
Chocolate Ice-Cream	●									
Chocolate Pops							●		●	
Chocolate Syrup	●									
Cocoanut Creams				●					●	
Cordials		●	●							
Culinary Products	●									
Frappe		●							●	
Fruit Flavored Creams				●					●	
Fudge	●	●	●		●			●		
Hand Rolls	●	●								
Ice-Cream Bars							●		●	
Italian Creams		●								
Jelly Centers				●					●	
Lotions										●
Marshmallows, Marshmallow Crackers		●	●	●	●			●		
Nougats			●	●	●	●		●	●	
Nuts, Nut Clusters, Nut Rolls				●	●	●		●		
Nut Flavored Creams			●	●					●	
Ointments										●
Peppermint Creams	●	●	●							
Raisins, Dates, Figs				●		●		●		
Sponge Centers				●					●	
Suppositories									●	●
Tobacco	●									
Vanilla-Creams				●		●		●	●	

APPROXIMATE COMPOSITION OF CHOCOLATE COATINGS

	Chocolate Liquor	Sugar	Total Fat	Whole Milk Solids	Skim Milk or Buttermilk Solids
Bitter..............	100	0	50–56	0	0
Bitter-Sweet........	80–95	5–20	40–50	0	0
Mild Bitter-Sweet...	50–75	25–38	38–42	0	0
Dark-Sweet.........	25–45	40–50	34–42	0	0
Medium Dark-Sweet	20–40	45–55	32–42	0	0
High Grade Lt.-Sweet	15–25	40–50	34–39	0	0
Low Grade Lt.-Sweet	5–15	55–68	31–39	0	0
Sweet Milk.........	7–17	35–55	28–39	12+	0
Ice-Cream Bar......	9–29	29–39	50–60	12+	0–6
Skim Milk, Buttermilk, Malted Milk	7–17	35–55	30–40	0	12+

Small quantities of flavoring materials such as vanillin, coumarin, coffee, cinnamon, peppermint, orange, lemon, anise, and synthetic cocoa-bean flavor also may be used.

COFFEE

The fruit of Coffea Arabica.

Coffee is indigenous to tropical Africa, more particularly to Abyssinia, from whence its growth undoubtedly spread across the continent to Liberia, where it grows wild and where it is also now extensively cultivated. The Liberian coffee plant is a distinct species, which seems to be less subject to disease and which has been successfully introduced into the East Indies.

The cultivation soon spread to the Western world, Cuba, Mexico, Central and South America, and at the present time Brazil produces about three-quarters of the whole world market.

In its wild state the coffee tree attains a height of from 20 to 30 feet and has somewhat the aspect of a cherry tree. In cultivation it is trimmed down to the height of about 6 feet. The coffee tree develops best in rich soil and in a tropical climate where the rainfall exceeds 75 inches per year.

The fruit is an oval, deep purple, two-seeded drupe, the parchment-like endocarp inclosing two plano-convex seeds, which are placed together by their flat sides and constitute the raw coffee of commerce. There are two general varieties:

BRAZIL COFFEES and MILD COFFEES

The Brazil coffees include the Santos and Rio coffees, the Mild coffees include the Java, Mocha, Bogota, Guatemala, Porto Rican, Mexican, Maracaibo, and a few others.

SANTOS COFFEE

Is shipped from the port of Santos. It is a mild and sweet coffee.

Santos coffee is usually blended with Bogota to obtain a good flavor and color.

RIO COFFEE

Is shipped from the port of Rio de Janeiro. It is a rather pungent coffee.

The roasting of coffee results in a notable reduction of some of the constituents, especially the caffein, fat, and sugar.

When properly conducted, the total loss in weight amounts to from 12 to 18%, of which about 8% represents moisture.

Coffee is sometimes glazed with sugar before roasting. According to Hilger and Juckenack, glazed coffee needs to be heated to a much higher temperature, which results in about double the loss of caffein and fat.

French roasted coffees belong to this class.

TEA

The tea plant is an evergreen shrub, which grows to a height of from 12 to 15 feet in the wild state, but under cultivation is kept about 4 feet high. Tea will grow in a variety of climates, but the sub-tropical appears to be the best, especially in sections where the rainfall is about 50 inches annually. According to the method of curing, three different kinds of tea are recognized:

I. Green or Unfermented Tea
II. Black or Fermented Tea
III. Oolong or Semi-Fermented Tea

All of these teas can be produced from leaves picked from the same bush.

GREEN TEAS

A. JAPAN GREEN TEAS
   1. Basket-Fired Green Tea (long leaves)
   2. Pan-Fired Green Tea (short leaves)

B. CHINA GREEN TEAS
   1. Gunpowders (small round rolled leaves)
   2. Imperials (large round rolled leaves)
   3. Young Hysons (long rolled leaves)

There is no difference in the drinking qualities of these subdivisions, only in appearance. CHINA GREEN TEAS are generally known as SOUCHONG.

OOLONG TEAS

Oolong teas are produced on the Island of Formosa. They are fired when the leaves are only half fermented.

BLACK TEAS

A. Black Teas produced in China: Called English Breakfast Tea or Congou Tea.

B. Black Teas produced in India, Ceylon, Java, and Sumatra are classified according to the size of the leaf. The three young shoots at the tip are known as ORANGE PEKOE and PEKOE, which contain the least fiber, most juice, and produce the finest grade of tea. The leaves immediately below the Pekoe are named SOUCHONG, a tea of popular price. Pekoe and Souchong are sometimes mixed and sold as PEKOE SOUCHONG.

The difference between green and black tea is mainly due to the fact that the green tea is dried shortly after gathering, and then rolled and carefully fired, whereas black tea is first made up into heaps, which are exposed to the air for some time before firing and allowed to undergo fermentation, resulting in the conversion of its original olive green color into a black color.

CHAPTER III

CANNED FRUITS, VEGETABLES, SOUPS, AND DRIED FRUITS

Canned foods have now come to occupy an important place in practically every home. The output of the canning industry is tremendous. Canned foods are prepared on a large scale, which tends toward economy.

The process of preserving food by means of heat alone was discovered by a Frenchman, Nicholas Appert. His method was to place the food, whether raw or cooked, in glass jars containing sufficient water to cover the food, then to place these jars in a vessel or bath of warm water, which was gradually heated until the temperature in the center of the jars reached 190–200° F. The jars were then corked and allowed to cool slowly in the water.

Canning was introduced into the United States by Ezra Daggett and Thomas Kensett, who canned salmon, lobster, and oysters in New York in 1819, and by William Underwood, of Boston, who, in 1820, put up some fruits, pickles, and condiments in glass, principally for the export trade. Little progress was made until 1840–1850, when fish canning was begun at Eastport, Maine, and oysters were canned in Baltimore. In 1845 the first canning company in Maine was organized. Ever since, the canning industry has advanced and progressed, chemically and bacteriologically, in leaps and bounds.

CANNED GOODS *

Canned goods are commonly packed in 5 sizes of cans, the commercial designation and average net contents of which are:

No. 1	12 oz.
No. 2	1 lb. 4 oz.
No. 2½	1 lb. 13 oz.
No. 3	2 lb. 3 oz.
No. 10	6 lb. 9 oz.

* Almanac of The Canning Industry, The Canning Trade, Baltimore, Md. Grading of Canned Fruits and Vegetables.

CANNED FRUITS

GENERAL DESCRIPTION OF GRADES

FANCY	Superlative quality. The top grade and packed in extra heavy syrup.
CHOICE	The grade between the top and the standard, packed in heavy syrup. Fine quality.
STANDARD	The medium grade, packed in medium syrup. Good quality.

Second Quality Fruit Packed

In 10% syrup to be designated:	"Second"
In water to be designated:	"Water"
For bakery use to be designated:	"Pie" and "Solid Pack Pie"

HEAVY SYRUPS To be used in Apricots, Peaches, and Plums and beginning with 10% sugar by weight on Seconds, to increase in steps of 15% to each grade. That is, for

SECONDS	10% sugar when packed
STANDARDS	25% sugar when packed
CHOICE	40% sugar when packed
FANCY	55% sugar when packed

LIGHT SYRUPS To be used on Pears, Cherries, and Grapes, and beginning with 10% sugar by weight on Seconds, to increase in steps of 10% to each grade. That is, for

SECONDS	10% sugar when packed
STANDARDS	20% sugar when packed
CHOICE	30% sugar when packed
FANCY	40% sugar when packed

Either the "BRIX" or "BALLING" scale is used on hydrometers to indicate the percentage by weight of sugar going into the solution. The terms "Brix" or "Balling" are the same when using the hydrometer.

UNLESS OTHERWISE STATED, FRUITS ARE UNDERSTOOD TO MEAN

APRICOTS	Approximate halves, unpeeled fruit
PEARS and PEACHES	Approximate halves, peeled fruit
CHERRIES, PLUMS, GRAPES	Whole unpeeled fruit

SPECIFICATIONS FOR CALIFORNIA CANNED FRUITS

MAXIMUM NUMBER OF PIECES PER CAN

	3" "8 oz."	3¼" "8 oz."	Picnic	#1 Flat	#1 Tall	#2 Tall	#2½	#3	#10
Apricots									
Fancy	9	10	12	14	16	19	24	29	86
Choice	11	12	15	18	20	24	30	36	108
Standard	13	15	18	23	26	31	42	50	151
***Pears and *Peaches**									
Fancy	4	5	6	7	8	9	12	14	43
Choice	5	6	7	9	10	12	15	18	54
Standard	7	7	9	11	13	15	21	25	76
Royal Anne Cherries									
Fancy	31	33	41	51	57	67	85	102	306
Choice	38	41	51	63	71	83	105	126	378
Standard	46	50	62	79	89	106	145	174	522
White Cherries and Black Cherries									
Fancy	37	40	49	60	67	79	100	120	360
Choice	46	50	61	75	84	99	125	150	450
Standard	56	61	75	95	107	129	175	210	630

* Pears and Peaches in the "8 oz." Cans (3" and 3¼") not otherwise specified may be either halves or quarters.

MINIMUM NUMBER OF PIECES PER CAN

All Grades	3" "8 oz."	3¼" "8 oz."	Picnic	#1 Flat	#1 Tall	#2 Tall	#2½	#3	#10
Apricots	(No minimum)								
Pears and Peaches	3	3	3	3	4	4	6	6	18
Royal Anne Cherries	(No minimum)								
White Cherries and Black Cherries	(No minimum)								

MAXIMUM VARIATION IN THE NUMBER OF PIECES PER CAN

	3" "8 oz."	3¼" "8 oz."	Picnic	#1 Flat	#1 Tall	#2 Tall	#2½	#3	#10
Apricots									
Fancy	3	3	3	4	4	5	6	7	21
Choice	3	3	3	4	5	6	7	8	25
Standard	4	4	4	5	6	7	8	9	29
Pears and Peaches									
Fancy	2	2	2	3	3	3	4	5	14
Choice	2	2	2	3	3	4	5	6	18
Standard	3	3	3	4	4	5	6	7	22
Royal Anne Cherries									
Fancy	6	6	8	10	12	14	16	19	58
Choice	8	8	10	12	15	17	20	24	72
Standard	16	16	20	25	30	35	40	48	144
White Cherries and Black Cherries									
Fancy	7	7	10	12	14	16	20	24	72
Choice	9	9	12	15	17	20	25	30	90
Standard	18	18	25	30	35	40	50	60	180

FANCY **In Syrup, 55% Sugar When Packed.**

Fruit to be of very high color, * ripe yet retaining its form and free from blemishes serious for the grade, halves uniform in size and very symmetrical.

CHOICE **In Syrup, 40% Sugar When Packed.**

Fruit to be of high color, * ripe yet retaining its form and free from blemishes serious for the grade, halves uniform in size and symmetrical.

STANDARD **In Syrup, 25% Sugar When Packed.**

Fruit to be of reasonably good color and reasonably free from blemishes serious for the grade, halves reasonably uniform in size, color, and degree of * ripeness, and reasonably symmetrical.

* Fruit, ripe to the point of being flattened or ragged, to be considered a proper delivery in all grades, the degree of raggedness permissible depending on the grade, being least (that is, fruit only slightly ragged being permitted) in the Fancy grade and most (that is, fruit distinctly ragged being permitted) in the Second grade.

Syrup made cloudy by particles of fruit, disintegrated on account of its ripeness, to be considered a proper delivery in all grades, the degree of cloudiness permissible depending on the grade, being least (that is, syrup only slightly cloudy being permitted) in the Fancy grade and most (that is, syrup distinctly cloudy being permitted) in the Second grade.

SECONDS **In Syrup, 10% Sugar When Packed.**

Trade designation "Seconds."

Fruit to be tolerably free from blemishes serious for the grade, halves tolerably uniform in size, color, and degree of * ripeness.

Packed in Water.

Trade designation "Water."

Fruit to be tolerably free from blemishes serious for the grade, halves tolerably uniform in size, color, and degree of * ripeness.

Packed for Bakery Use.

Wholesome fruit unsuited for above grades.

Trade designation "Pie": Minimum drained weight 70 ounces.

Trade designation "Solid Pack Pie": Minimum drained weight 92 ounces.

* See p. 121.

CANNED FRUITS

CHERRY

About 250 varieties of cherries are grown.

OXHEARTS (black and white) TARTARIAN BIGERAN	Are best for eating
MORELLO MONTMORENCY EARLY RICHMOND	Are cooking cherries (for pies, puddings)

ROYAL ANNE cherries are considered the best canned cherries.

APRICOTS

Closely related to the peach, and originally came from Armenia. They are grown extensively in Japan and China, and also in California. The kernel is used in making "Kernel Paste," a substitute for almond paste.

Grades	PARK	A fleshy fruit
	BLENHEIM	High in color

Apricots are sold in canned form and as dried fruit. The so-called Spanish apricot has a higher acidity than the domestic and the latter is therefore preferred in preparations where an excess of acid would be harmful (See Apritop).

PEARS

Although a great many different pears are grown in America and Europe, the BARTLETT and KIEFFER varieties are commonly found in the canning industry.

PEACHES

Originally grown in China, are now cultivated in Europe, Africa, Asia, and also in America. In general there are two kinds of peaches

FREE, in which the pulp separates readily from the stone, and

CLINGS, in which the pulp adheres strongly to the stone.

Peaches are dried and canned. The canned peaches are packed under the following trade designations:

YELLOW CLING PEACHES	Fine appearance and color
SLICED YELLOW CLING PEACHES	
YELLOW FREE PEACHES	Inferior in appearance but of peculiar, much preferred flavor

The dried peaches are of the yellow free variety.

NECTARINE

A variety of the peach. Smaller in size and with a smooth skin.

MUSCAT GRAPES

A very sweet, spicy-tasting grape.

PLUMS

The plum is a native of Asia, brought to Europe 100 years ago, and shortly after to America.

Varieties
- The Cherry plum in Europe
- The Japanese plum
- The Chinese plum
- The Wild Goose plum in America
- The Sand plum in America
- The Damsons in England
- The Green Gage in England
- The Mirabelles in England

PRUNES

The fresh prunes usually canned are the OREGON and the ITALIAN varieties.

The syrup must test not less than 20% sugar when packed. The minimum drained weights thirty days after packing to be as follows:

3" "Eight Ounce"	6¼ oz.	No. 2 Tall	15 oz.
3¼" "Eight Ounce"	6¾ oz.	No. 2½	22 oz.
Picnic	8 oz.	No. 10	79 oz.
No. 1 Tall	12 oz.		

At time of packing the weight usually runs 4% to 5% less than the above and increases gradually for thirty days or longer.

FRUITS FOR CANNED FRUIT SALADS

APRICOTS	Halves
BARTLETT PEARS	Halves or quarters
YELLOW CLING PEACHES	Halves, quarters, sixths, or eighths
PINEAPPLE	Sectors
MARASCHINO TYPE CHERRIES	White, pitted, colored red with U. S. food colors

APPLES

The tins shall be well filled with ripe, firm winter apples in water.

APPLE BUTTER

Apple butter shall be made from whole, sound, fresh, ripe fruit, with no cores, peelings, or stems, free from preservatives and glucose. The finished product shall contain not less than 55% of solids.

A. CANNED SOUPS

1. Consomme

   A clear soup, made from beef broth and a combination of the juices of several vegetables, usually those of carrots, turnips, onions, leeks, and parsley.

2. Chicken Gumbo

   A soup made from chicken and tomato stock, to which is added chicken meat, rice, and okra (gumbo). The latter gives the soup its peculiar flavor.

3. Julienne

   A soup made from beef broth, small peas, shredded carrots, turnips, celery, leeks, and cabbages. A medium thick soup.

4. Mock Turtle

   A dark colored, rich, thick soup. Made from tomato and meat juice, thickened with flour, and flavored with spices and herbs.

5. Tomato Soups

Tomato Bouillon	Made from tomato juice and beef broth.
Tomato Soup	Cooked down tomato juice, flavored with salt, sugar and spices. Thickened with flour.
Tomato Cream Soup	Made from tomato juice and added cream.

6. Mulligatawny

   Meaning "pepper water." A hot and spicy soup, made from chicken stock and ground tomatoes, flavored with curry powder.

B. CANNED VEGETABLES

BEANS

State Beans	Pea bean.
California Navy Beans	In "small" and "large" size.
Michigan Navy Beans	An easy and quick-cooking bean.
Kintoki Beans	From China and Japan; used principally in chili con carne.
Pinto Beans	A freely speckled bean. Very tender.

Beans are put up in four different ways:

Beans with Plain Sauce	Molasses, sugar, salt, water.
Beans with Tomato Sauce	Molasses, sugar, salt, tomato pulp, spices.
Beans with Pork and Plain Sauce	Slices of pork and plain sauce.
Beans with Pork and Tomato Sauce	Slices of pork and tomato sauce.

STRING BEANS

There are two different kinds of string beans.

The flat pod bean and
The round or refugee bean.

Both can be wax or green beans.

No. 2 can shall contain not less than 13 ounces after draining.

No. 10 can shall contain not less than 4 pounds after draining.

BEAN SPROUTS

These are sprouted beans of the MUNG variety, a bean indigenous to China, but now also grown in the United States. It is a small, round bean.

"Bean sprouts" are used in chop suey and other Chinese food preparations.

LIMA BEANS

Lima beans, if canned, shall be medium sized, fresh green or white limas, free from undeveloped, overripe, or hard beans.

No. 2 tins shall contain not less than 13 ounces after draining.

No. 10 tins shall contain not less than 4 pounds after draining.

PEAS

There are two kinds of peas:

1. The early June peas, which mature in June. They are round, of smooth skin, and are the more expensive kind.
2. The sweet peas, which mature in July. They are irregular in shape, mostly oblong, and have a wrinkled skin.

The peas themselves are distinguished by their size. According to the mesh of the screen through which they pass, they are classified as:

No. 1; No. 2; No. 3; No. 4; No. 5; No. 6

Peas are put up in a brine consisting of water, sugar, and salt.

A No. 2 can shall contain not less than 12 ounces of peas after draining.

A No. 10 can shall contain not less than 4 pounds 8 ounces of peas after draining.

"Alaska" and "Winner" peas are trade names for early June peas, not indicating their origin.

ASPARAGUS

There are three kinds of asparagus:

The white asparagus—the spear is kept under the ground.

The green asparagus—the spear grows above the ground.

The peeled asparagus—the spear is freed from its skin.

There shall be about 40 stalks per No. 2½ can.

CARROTS

Carrots are put up in cubes with a brine like peas, and together with peas under the name of Carrots and Peas.

OKRA

Okra is the young pod of the okra or gumbo plant.

The pods have from five to twelve sides and are from two to ten inches long. They contain numerous small seeds.

Okra is used principally in gumbo soup.

DRIED BEANS

Baby Lima	Ceci
Black Eye	Fava (a large bean)
Black Turtle	Lima

Grades: Extra Choice, Recleaned, Choice Recleaned, Sample.

Marrow	Roman
Pea	Soya
Red Kidney	Yellow Eye

TOMATOES

Canned tomatoes must be clean, sound, ripe, fairly evenly colored, well-peeled, and cored. The can, upon being opened and drained over a sieve of ¼-inch mesh for 2 minutes, shall show not less than 50% of the contents above the screen.

TOMATO PUREE

May be made from either strained tomatoes * or trimming stock,** but when made from the latter the label must carry an explanatory legend equivalent to "Made from trimmings."

The customary requirements for tomato puree are:

Specific Gravity for Fancy Goods 1.045
Specific Gravity for Heavy Goods 1.050

SUCCOTASH

Succotash is a mixture of 75% corn and 25% lima beans.

CORN

Canned corn must be well cleaned and must be free from silk, bits of cob, and other foreign material. It should be of New York, Maryland or Indiana grading.

A brine consisting of salt, sugar, and water is added.

"SHOE-PEG" Corn, is corn on the cob minus the cob.

No. 2 tins shall contain not less than 13.5 ounces of corn after draining.

No. 10 tins shall contain not less than 4 pounds 8 ounces after draining.

SPINACH

Canned spinach must be solid-packed, from fresh green stock, and must be free from grit. No. 3 shall contain not less than 1 pound, 9.5 ounces after draining the tin for 2 minutes over a ⅛-inch mesh screen.

PIMENTOS

Pimentos are large, sweet, red peppers that have been peeled and put up in their own juice.

B. DRIED FRUITS

APRICOTS

See under Canned Fruits.

CITRON

The citron grows on a tree similar to an orange or lemon tree, which is cultivated in China, Persia, Greece, Palestine, Italy, and California.

The fruit is an oblong, 5–9 inches long, and resembles a lemon, but has a very thick peel and a smaller amount of acid pulp.

The peel, thick and spongy, is used in the preparation of CANDIED or PRESERVED CITRON, but is also used in the dried state as

* Strained tomatoes. Obtained by straining sound, ripe tomatoes, either raw or cooked, through a screen that removes skins and seeds.

** Obtained by straining sound peeling, pieces and trimmings from ripe tomatoes through a screen that removes skins and seeds.

CITRON PEEL.

The drained citron peel is sugared, but is not coated with sugar, as the candied kind is.

DATES

Mesopotamia produces about 80% of the world supply of dates, the fruit of the date palm. The palm is cultivated in Western Asia and Northern Africa, also in Southern California and Arizona. The fruit grows in bunches of 20 pounds and more.

Varieties

The Hallawi Date	From Persia. Firm in texture, bright golden in color.
The Khadrawi Date	Also from Persia. Less firm in texture, darker in color.

Both are alike in flavor.

Dates contain about 55% of sugar and 7% of protein; they are of high nutritive value.

Fard Date, a dark colored, firm Arabian date.

FIGS

The fig is largely cultivated in the Mediterranean countries. It appears like a pear when fresh. The harvesting time is in June and again later in fall. Only the latter can be dried and shipped.

Varieties

Smyrna Figs	Large, sweet, white fruits. Shipped in small cases, boxes, or baskets, pressed flat. They are considered the best.
Adriatic Figs	Shipped from Greece, squeezed together and placed on strings.
Calimyrna Figs	Grown in California from stock brought from Smyrna, Turkey.
Istrian Figs	Small, very sweet. Do not keep well. Shipped mostly in barrels or cases.
Mission Black	A black fig, cultivated in the Southern States.

PRUNES

Are dried plums. (See under Canned Fruit.)

California takes the foremost place in producing prunes.

There are three ways of curing prunes:

I. Sun-dried
II. Curing in evaporators.
III. Partially cooking and then drying either by method I or II.

Varieties

California or French Prunes	Very sweet
Oregon or Italian Prunes	Tart
Oregon Petites	Sweeter than Oregon

Prunes are graded into ten sizes:

20– 30s	To the pound
30– 40s	
40– 50s	
50– 60s	
60– 70s	
70– 80s	
80– 90s	
90–100s	
100–110s	
110–120s	

RAISINS

Are dried grapes.

There are several special grapes, which are best adapted for making raisins:

I. The MUSCATELS or sun-dried raisins. From MALAGA and VALENCIA.

II. The VALENCIAS or dipped raisins. Dipped in potash lye, containing lavender and rosemary, and in olive oil.
From Valencia, called Denis.
From Turkey, called Smyrna.
From California, called Valencias.

III. The SULTANAS, the small, tart, seedless raisins from Turkey, Greece, and Persia.

IV. The THOMPSON RAISIN—a true seedless, very sweet raisin from California.

V. The "PUFFED" RAISINS—seedless raisins, not sticky.

Size Grades

1 Crown	Raisins that at time of boxing will not pass through a screen with opening smaller than 12/32 inch.
2 Crown	Raisins that at time of boxing will not pass through a screen with opening smaller than 17/32 inch.
3 Crown	Raisins that at time of boxing will not pass through a screen with opening smaller than 21/32 inch.
4 Crown	Raisins that at time of boxing will not pass through a screen with opening smaller than 21/32 inch.
5 Crown, 6 Crown	Hand sorted jumbo, generally sold in clusters.

CURRANTS

The dried fruit of the small black CORINTH grape in Greece. The currant growing in the United States is the fruit of a bush like the gooseberry and is a different fruit altogether.

Grades
- Vostizza
- Patras
- Amalias
- Pyrgos

Patras and Amalias are most commonly imported in the U. S. A.

TAMARINDS

The fruits of a tree cultivated in the East and West Indies and in Africa.

The pulp found in commerce is mostly used in medicine as a purgative. But the raw fruits are also used in the manufacturing of chewing tobacco and the extract prepared from the fruits is often added to beverages, where a tart taste is preferred. The East Indian tamarinds are considered the best.

CHAPTER IV

FOUNTAIN SUPPLIES, CRUSHED FRUITS, FLAVORS, EXTRACTS, ESSENCES, EMULSIONS, CANDY, CHEWING GUM, FOOD COLORS, BEVERAGES, BITTERS, CORDIALS, VERMOUTH, COCKTAILS, AND SYRUP TABLES

Fountain supplies are a class of products used at soda fountains for carbonated beverages, ice-cream, sodas, and household purposes. Crushed fruits, flavors, extracts, etc., are employed in such fields as the ice-cream industry, candy making, bottlers of soft drinks, household extracts and flavors, etc.

The manufacturer of these products should exercise care in selection of ingredients, cleanliness in his manufacturing rooms, accuracy in his calculations, and purity of his products. Also remember that the finished product must be stable, and withstand climatic changes, and must not ferment or become mouldy.

The small manufacturer can almost in every case build home-made apparatus. In starting a new business, if you must purchase equipment, arrange to get some credit for it, in exchange for larger equipment later on.

FLAVORING EXTRACTS

ALMOND EXTRACT—60% Alcohol

Bitter Almond Oil (free from Prussic Acid)	4	oz.
Alcohol	4¾	pt.
Cherry Laurel Water	6	oz.
Water for	1	gal.

ALLSPICE EXTRACT—50% Alcohol

Pimento Oil	1	oz.
Alcohol	8	oz.
Water for	1	pt.

ANISE EXTRACT—72% Alcohol

Anise Oil	4	oz.
Alcohol	5	pt. 12 oz.
Balance Water for	1	gal.

Or

Anise Oil	4	oz.	
Caraway Oil	1/4	oz.	
Glycerin	4	oz.	
Alcohol	5	pt.	6 oz.
Water for	1	gal.	

(N. B. use for Anisette—1 ounce to the gallon.)

BERGAMOT EXTRACT—60% Alcohol

Bergamot Oil	3	dr.
Pineapple Ether	2	oz.
Alcohol	10	oz.
Water	4	oz.

CARDAMOM EXTRACT—37.5% Alcohol

Cardamom Oil	1	oz.
Alcohol	3	pt.
Water	5	pt.

Or

Cardamom Seeds	1	oz.
Alcohol	12	oz.
Water	4	oz.

Macerate 8 days.

CASSIA EXTRACT—66% Alcohol

Cassia Oil	8	oz.
Alcohol	5 1/4	pt.
Water for	1	gal.

CARAWAY EXTRACT—75% Alcohol

Caraway Oil	8	oz.
Anise Oil	1/2	oz.
Glycerine	8	oz.
Alcohol	6	pt.
Water for	1	gal.

CELERY EXTRACT—75% Alcohol

Celery Oil	1 1/2	oz.
Alcohol	12	oz.
Water	4	oz.

CINCHONA EXTRACT (Peruvian Bark Extract)

Yellow Cinchona Bark Powder	1	lb.
Glycerin	4	oz.
Alcohol and Water for	16	oz.

CINNAMON EXTRACT—66% Alcohol

Cinnamon Oil	6	oz.
Alcohol	5¼	pt.
Water for	1	gal.

COFFEE EXTRACT

A. EXTRACT MADE BY MACERATION:

Coffee, roasted and ground	15	lb.
Alcohol	2½	gal.
Water	2½	gal.

Macerate 8 days.

B. EXTRACT MADE BY INFUSION:

Coffee, roasted and ground	5	lb.
Alcohol	1	qt.
Water	1	qt.
Syrup	1	qt.

C. A NEW AND BETTER EXTRACT:

Ground Coffee	1	lb.
Glycerin	1	pt.
Water	1	pt.

Boil the coffee in a bag with the glycerin and water until all the water is expelled.

Digest for about 1 week:

Coffee	1	lb.
Water	1	pt.
Alcohol	1	pt.

Add the two together and add a trace of lime juice to prevent the coffee from getting stale.

D. EXTRACT MADE WITH SOLUBLE COFFEE:

Soak 100 lb. coffee in 40 gal. hot water.

Percolate (excelsior on the bottom) for 2 days.

Wash with hot water until 40 gal. are obtained.

Pack 100 lb. of coffee more and percolate for 2 days with the first menstruum.

Make up to 40 gal.

Dissolve ½ lb. SOLUBLE COFFEE to 1 gal. EXTRACT.

Alcohol 46%.

CHOCOLATE EXTRACT—48% Alcohol

Cocoa	15	lb.
Alcohol	2½	gal.
Water	2½	gal.

Let macerate for 1 month. Then filter.

Or

Cocoa or Chocolate	8	oz.
Balsam Peru Tincture	1	dr.
Alcohol	8	oz.
Water	8	oz.

COCOA EXTRACT

As Chocolate Extract.

FOENUGREEK EXTRACT—70% Alcohol

Foenugreek Seeds	8	lb.
Alcohol	2¾	gal.
Water	1¼	gal.

Macerate for 1 month. Then filter.

GINGER EXTRACT

Oleo-Resin Ginger	4	oz.
Add Alcohol	5¾	pt.

Dissolve. If not clear, filter.

Heat ½ gallon water to 60° C. and dissolve in it ½ ounce SODIUM CARBONATE ANHYDROUS C.P. Add this solution to the alcoholic Oleo-Resin Ginger Extract. Shake well and let stand 2 days. Then add 1 ounce Kaolin and shake well. Let stand for 2 weeks, giving the bottle a good shake each day. After 2 weeks decant and filter.

GRENADINE EXTRACT

Press full-ripe pomegranates. To the press-residue add 3 times the amount of alcohol and distill. The distillate can be improved by the addition of Grenadine Fruit Oil.

FERNET EXTRACT

Licorice Root	100	g.
Licorice Sweet	50	g.
Galanga Root	20	g.
Aloe	10	g.
Nutmeg	10	g.
Saffron	5	g.
Alcohol	¾	liter
Water	½	liter

Macerate.

LEMON EXTRACT—87½% Alcohol

Lemon Oil	32	oz.
Alcohol	4½	gal.
Water	1	qt.
Liquid Lemon Color *	5	oz.

* Made by dissolving 1 dram Tartrazin in 16 ounces water (Called TARTRAZIN SOLUTION).

Cut the oil with the alcohol, add the water slowly and color. Shake well and let stand. The extract becomes stronger on standing. If cloudy, filter through magnesia.

LEMON EXTRACT—59% Alcohol

Lemon Oil	40 oz.
Alcohol	3 gal. 1 pt.
Bitter Almond Oil	10 drops
Liquid Yellow Color	5 drops
Citric Acid	5 oz.
Water	1½ gal.

Cut the oils with the alcohol. Dissolve the citric acid in the water and add to the alcoholic oil mixture. Proceed as under 87½% Lemon Extract.

LEMON EXTRACT—18% Alcohol

Lemon Oil	15 oz.
Orange Oil	2½ oz.
Citric Acid	2½ oz.
Alcohol	7½ pt.
Water	3 gal. 7¼ pt.
Liquid Yellow Color	5 drops
Or	
Lemon Oil	20 oz.
Alcohol	7½ pt.
Color	5 drops
Water	3 gal. 7¼ pt.

Compound as under preceding formulae.

LEMON SUBSTITUTE—12½% Alcohol

Lemon Oil	2 oz.
Citral	¼ oz.
Alcohol	1 pt.
Water and Color for	1 gal.

LEMON FLAVOR—6% Alcohol

Terpeneless Lemon Oil	2½ dr.
Citral	2½ dr.
Lemon Oil	5 oz.
Alcohol	2½ pt.
Citric Acid	2½ oz.
Magnesia Carbonate	10 oz.
Liquid Color *	1½ oz.

* Made by dissolving 1 Dram Tartrazin in 16 ounces water (TARTRAZIN SOLUTION).

LEMON CITRAL FLAVOR—25% Alcohol

Lemon Citral **	2½	pt.
Alcohol	1	gal. 1 qt.
Lemon Color Liquid	5	oz.
Balance Water for	5	gal.

CITRAL LEMOLA

Lemon Citral **	6	oz.
Mazola Oil	1	gal.

LEMON CITRAL **

Half Lemon Oil
Half Citral

LEMON FLAVOR (Non-alcoholic)

Terpeneless Lemon Oil	3	oz.
Lemon Oil	5	pt. 13 oz.
Glycerin	1	gal. 7 pt.
Indian Gum	9	oz.
Winter Wheat Flour	9	oz.
Water	2	gal. 1 qt.

Mix the oils, add gum and flour and some water, shake well. Then add the glycerin, shake, and add the rest of the water. Shake until the whole forms a thin emulsion-like liquid.

CONCENTRATED LEMON EXTRACT (Terpeneless)

Lemon Oil, Terpeneless	1½	oz.
Alcohol	15	oz.

(Absolutely soluble in all proportions.)

TERPENELESS LEMON EXTRACT

(Terpeneless Lemon Extract must contain 0.2% citral.)

Terpeneless Lemon Oil (45% citral)	5	dr.
Alcohol and Water to	1	gal.

LIME EXTRACT—87½% Alcohol

Lime Oil, redistilled	24	oz.
Alcohol	2	gal. 5 pt.
Water	1½	pt.

MACE EXTRACT—65% Alcohol

Mace Oil	8	oz.
Alcohol	5¼	pt.
Water for	1	gal.

NUTMEG EXTRACT—65% Alcohol

(This is also sold as MACE EXTRACT. The mace oil being too expensive.)

Nutmeg Oil	8 oz.
Alcohol	5¼ pt.
Water for	1 gal.

MAPLE EXTRACT (Like Mapleine)

Coffee Extract	20 gal.
Foenugreek Extract *	108 oz.
Lovage Oil 1 : 10	5 oz.
Vanillin	13 oz.
Coumarin	1 oz.
Alcohol	4½ gal.
Hot Water	8 gal.
Caramel	10 gal.

MAPLE EXTRACT

Alcohol	20 oz.
Caramel	20 oz.
Vanillin	2.7 oz.
Lovage Extract	1.3 oz.
Foenugreek Extract	15.5 oz.
Peru Balsam	0.2 oz.
Coffee	16.8 oz.
Hickory Bark	5.5 oz.
Water	20 oz.

MAPLE EXTRACT (For compounding flavors)

Foenugreek Seeds, powdered	1 lb.
Lovage Root, ground	¼ lb.
Alcohol	½ gal.

Let stand a few days. Strain.

Coffee, pulverized	2 lb.
Boiling Water	1 gal.

Boil for 10 minutes. Cool and strain.

Dissolve

Vanillin	2 oz.
Rue Oil	1 dr.
Celery Oil	1 dr.
Glycerin	1 pt.

* Foenugreek Seeds ground	3¼ lb.
Alcohol	7 pt. (exact: 108 oz.)
Water	3 pt. (exact: 54 oz.)

Mix all these ingredients together with 2 pints HEAVY SYRUP or GLUCOSE, add enough water to make 2 gallons.

ONION EXTRACT

This extract is made from well-peeled onions which must be finely cut up. In order that the extract should not be too mucous in character, the onions are first well expressed, the juice being discarded. To every pound of the residue add 2 pounds alcohol. Let stand 3 days and then express. Treat the residue once again but with 2 pounds of 50% alcohol. Then express after 3 days and unite the liquids. To each pound of extract add ¼ dram true volatile mustard oil.

ORANGE EXTRACT—87½% Alcohol

Orange Oil	32	oz.
Alcohol	4½	gal.
Liquid Orange Color	2	oz.
Water	32	oz.

(To make an orange extract cloudy, add a few drops of Tincture of Benzoin.)

ORANGE EXTRACT—59% Alcohol

Orange Oil	8	oz.
Alcohol	5	pt.
Bitter Almond Oil	2	drops
Red Color Liquid	1	drop
Water for	1	gal.

ORANGE EXTRACT—18% Alcohol

Orange Oil	3	oz.
Lemon Oil	½	oz.
Citric Acid	½	oz.
Alcohol	24	oz.
Red Color Liquid	1	drop
Water for	1	gal.

PEPPERMINT EXTRACT—78% Alcohol

Peppermint Oil	8	oz.
Alcohol	6	pt. 4 oz.
Water for	1	gal.

Filter through magnesia.

SPICE EXTRACT—50% Alcohol

Black Pepper	1	lb.
Turmeric	12	oz.
Coriander	6	dr.
Pimento	12	dr.
Nutmeg	12	dr.
Cloves	12	dr.
Cinnamon	12	dr.
Caraway	12	dr.
Alcohol	1	qt.
Water	1	qt.

Macerate for 2 weeks. Decant and filter.

Or

Black Pepper	1	oz.
Allspice	½	oz.
Nutmeg	¼	oz.

Infuse in 16 fluid ounces brandy for 10 days.

ROOTBEER EXTRACT

Ginger Extract	12	oz.
Vanilla Extract	12	oz.
Sassafras Oil	4	oz.
Wintergreen Oil	2	oz.
Anise Oil	1	oz.
Orange Oil	¼	oz.
Clove Oil	¼	oz.
Alcohol	½	gal.
Simple Syrup	3½	gal.
Soap Bark Tincture	4	oz.
Salicylic Acid	1	dr.
Caramel	1¼	gal.
Water for	6	gal.

SPEARMINT EXTRACT—78% Alcohol

Spearmint Oil	8	oz.
Alcohol	6	pt. 4 oz.
Water for	1	gal.

TONKA EXTRACT—50% Alcohol

Tonka Beans, cut	50	lb.
Alcohol	25	gal.
Water	25	gal.

Macerate for 1 month.

SWEET BIRCH EXTRACT

Birch Oil	1½	oz.
Alcohol	12	oz.
Water	4	oz.

VANILLA EXTRACTS

PURE VANILLA EXTRACT—39% Alcohol

Mexican Beans, cut	Place in a bag	10	lb.
Bourbon Beans, cut		50	lb.
Alcohol		30	gal.
Water		30	gal.

Extract in a continuous-running percolator at a temperature of 110° F. for 48 hours.

To the extract add:

Sugar	50	lb.

dissolved in 4 gallons water.

The combination of the vanilla beans can be varied. Either the above proportions or a mixture of:

Bourbon Beans	40	lb.
Mexican Beans	5	lb.
Tahiti Beans	5	lb.

give good results (see note).

NOTE. If a dark extract is wanted add ½ pound BICARBONATE OF SODA. Put over the sift a CANTON FLANNEL (Felt), fuzzy side to the beans. Heat for best results to 140° F. for 48 hours.

VANILLA EXTRACT (By Maceration)—48% Alcohol

Vanilla Beans, cut	1	lb.
Alcohol	5	pt.
Water	3	pt.

Let stand 72 hours.

VANILLA EXTRACT (With Glycerin)

Run 4 ounces pure vanilla beans through a meat grinder. Add to this 1 pint glycerin and 1 pint water, previously mixed, and let stand for at least 2 weeks. Press and strain through muslin and add sufficient glycerin to make a total of 3 pints.

VANILLA EXTRACT—50% Alcohol

Standard Formula:

Vanilla Beans (Cut very fine)	14	oz.
Cologne Spirit, 190 proof	½	gal.
Distilled Water	½	gal.
Cane Sugar	8	oz.

VANILLA EXTRACT (With Licorice)

Cut ½ pound good vanilla beans in short pieces, place in a mortar with some fragments of broken glass and pound until fine. Place in a stone jug, or glass bottle, and add 1 ounce CORIGLIANO LICORICE in small pieces, 1 pint glycerin. ½ gallon alcohol, and 3 pints of water. Soak 3 weeks. Shake occasionally. Then filter.

VANILLA and TONKA EXTRACT—55% Alcohol

Add 1 quart PURE VANILLA EXTRACT to ¾ gallon TONKA EXTRACT.

VANILLA and TONKA EXTRACT

Vanilla Beans, Cut	4	oz.
Tonka Beans, Cut	2	oz.
Glycerin	8	oz.
Alcohol	3	pt.
Water	1½	pt.

Macerate 8 days and filter.

Or

Vanilla Beans, Cut	1¼	lb.
Tonka Beans, Cut	2	lb.
Alcohol	5	gal.
Water	3	gal.

Macerate 8 days and filter.

PURE VANILLA FLAVOR

Oleo-Resin Vanilla	4	oz.
Alcohol	2	pt.
Water to make	1	gal.

Flavoring ingredients must be completely dissolved in the alcohol before any water is added. Filter clear after 2 or 3 days.

VANILLA FLAVORS

COMPOUNDED

A. PURE VANILLA EXTRACTS (Reinforced with Vanillin and Coumarin)

Formula No. 1—35% Alcohol

Pure Vanilla Extract	½	gal.
Vanillin	4	oz.
Coumarin	2	oz.
Prune Juice	8	oz.
Glycerin	3	pt.

No. 2—46% Alcohol

Vanol	½ gal.
Pure Vanilla Extract	1 qt.
Water	1 qt.

No. 3

Pure Vanilla Extract	7½ pt.
Vanol	½ pt.

No. 4

Pure Vanilla Extract	1 pt.
Vanol	5 oz.
Vanilla Flavor Compound Formula No. 1	7 pt.

No. 5—25% Alcohol

Pure Vanilla Extract	1 pt.
Vanol	4 oz.
Select Vanilla	108 oz.

No. 6—52% Alcohol

Pure Vanilla Extract	¾ gal.
Vanilla Flavor Compound No. 1	1 qt.
Vanol	2 oz.

B. VERY CONCENTRATED VANILLA COMPOUNDS

One pint of these vanilla compounds mixed with warm water will produce 2 gallons of an ordinary vanilla extract of good strength.

MEXOLA TYPE—30% Alcohol

Vanillin	5¾ lb.
Coumarin	1 lb. 6 oz.
Alcohol	3 gal. 1 pt.
Glycerin	7 gal.
Sugar Color (Caramel)	4 gal.

VANOL, VANIL TYPE—25% Alcohol

Vanillin	3 lb.
Coumarin	1 lb.
Alcohol	1½ gal.
Glycerin	3¾ gal.
Caramel	¾ gal.

WHITE VANOL, VANIL TYPE

Vanillin	3 lb.
Coumarin	1 lb.
Alcohol	1½ gal.
Glycerin	4 gal.

VANOL, VANIL (N.A.C. = Without Alcohol) TYPE

Vanillin	30	oz.
Coumarin	10	oz.
Glycerin	4	gal. 8 oz.
Caramel	7	pt. 8 oz.

WHITE VANOL, VANIL N.A.C.

The same, leaving out the Caramel.

VANOLA—12% Alcohol

Vanillin	40	oz.
Coumarin	20	oz.
Glycerin	4	gal. 3 pt.
Alcohol	5	pt.
Sugar Color	20	oz.

C. VANILLA FLAVORS (Compounds)

Formula No. 1—21% Alcohol

A very popular vanilla flavor used successfully by bakers, confectioners, pastry chefs, ice-cream makers, etc.

Vanillin	4	lb. 6 oz.
Coumarin	1	lb.
Alcohol	11	gal.
Sugar	40	lb.
Caramel	½	gal.
Water	39	gal.

Dissolve the vanillin and coumarin in the alcohol, the sugar and caramel in the water, and mix both together.

No. 2—15% Alcohol

Vanillin	3	lb.
Coumarin	12	oz.
Alcohol	8	gal.
Sugar	40	lb.
Caramel	½	gal.
Water	41	gal.

Proceed as under No. 1.

No. 3

Vanilla No. 2	1	gal.
Vanol	1	oz.

No. 4—14% Alcohol

Vanilla No. 2	5	pt.
N.A.C. Vanilla	3	pt.

No. 5—9½% Alcohol

Vanillin	2½ lb.
Coumarin	6 oz.
Syrup	3 gal.
Glycerin	3 gal.
Alcohol	4 gal. 1 qt.
Sugar Color (Caramel)	12 oz.
Balance Water for	50 gal.

No. 6—6% Alcohol

Vanillin	5 oz.
Coumarin	1 oz. 2 dr.
Alcohol	5 pt.
Syrup	10 pt.
Caramel	10 oz.
Water for	10 gal.

N.A.C. VANILLA (Without Alcohol)

Vanillin	3 lb.
Coumarin	1½ lb.
Glycerin	6½ gal.
Caramel	4 pt.
Water for	50 gal.

HOUSEHOLD VANILLA (No Alcohol)

Vanillin	1 lb. 9 oz.
Coumarin	12½ oz.
Glycerin	3 gal. 1 pt.
Caramel	2 pt. 8 oz.
Water for	50 gal.

WHITE VANILLA—41% Alcohol

Vanillin	2 oz.
Coumarin	½ oz.
Alcohol	2 pt.
Syrup	1 pt.
Balance Water for	1 gal.

Or

Vanillin	10 oz.
Coumarin	2½ oz.
Alcohol	1 gal. 2 pt.
Syrup	5 pt.
Balance Water	3 gal. 1 pt. for 5 gal.

DIFFERENT FORMULAE

IMITATION VANILLA

Standard Formula:

Vanillin	7 oz.
Coumarin	1 oz.
Cane Sugar	8 lb.
Alcohol	3 gal.
Water enough for	10 gal.
Caramel Color	To suit

IMITATION VANILLA FLAVOR

(Stronger, and cheaper, but lacking the delicate character of the former.)

Vanillin	4 oz.
Coumarin	1 oz.
Cane Sugar	6 lb.
Alcohol	½ gal.
Water enough for	7 gal.
Caramel Color	To suit

GERMAN SYNTHETIC VANILLA EXTRACT

Vanillin	30 g.
Coumarin	0.33 g.
Alcohol	150 cc.
Glycerin	180 cc.
Caramel	40 cc.
Water enough for	4000 cc.

VANILLA SUBSTITUTE—19% Alcohol (Very strong flavor)

Vanillin	4 oz.
Coumarin	4 oz.
Alcohol	1 gal.
Syrup	½ gal.
Water	3½ gal.
Little Caramel Color	

VANILLA SUBSTITUTE—9½% Alcohol

Vanillin	3 oz.
Coumarin	1½ oz.
Alcohol	½ gal.
Syrup	½ gal.
Water for	5 gal.
Very little Caramel Color	
Eugenol	1 drop

VANILLA SUBSTITUTE—7% Alcohol

Vanillin	1 oz.
Coumarin	1 oz.
Syrup	½ gal.
Alcohol	3 pt.
Balance Water for	5 gal.
Very little Caramel Color	

VANILLA SUBSTITUTE—6% Alcohol

Vanillin	2½ oz.	Alcohol	2½ pt.
Coumarin	⅝ oz.	Sugar Color	5 oz.
Sugar	5 lb.	Water for	5 gal.

Or

Vanillin	1 lb.	Alcohol	1½ gal.
Coumarin	7 oz.	Sugar Color	25 oz.
Sugar	25 lb.	Water for	25 gal.

This last formula makes a cheap vanilla flavor, largely sold to bakers, etc.

FOAM PRODUCERS

SOAP BARK FOAMS

Formula A

Quillaja bark is used in the form of tincture and may be prepared as follows:

Quillaja, fine chips	5½ av. oz.
Alcohol	10 fl. oz.
Water	Sufficient

Mix the drug with 24 fluid ounces of water, boil for 15 minutes. Strain and add enough water through the strainer to make the volume equal to 22 fluid ounces. Mix the liquid when cool with the alcohol, let stand for 12 hours, filter, and to the filtrate add enough water to measure 32 fluid ounces.

If a cheaper preparation is desired, the alcohol may be replaced by water or by glycerine. If the former be used, the preparation must be preserved by the addition of a small amount of salicylic acid solution. Either of the latter is to be preferred to the alcoholic solution, as the alcohol has the tendency to cause premature expulsion of gas from the soda when served.

About one fluid ounce of this preparation is usually sufficient for one gallon of syrup.

Formula B

Soap Bark (chips)	1 lb.
Boiling Water	10 pt.
Alcohol (95%)	1 pt.

Boil the soap bark in the water for 30 minutes. Allow to cool. Add the alcohol. Pack a small quantity of dry soap bark in a percolator to make a bed and percolate. One-half to 1 ounce of this is used per gallon of syrup.

GINGER ALE EXTRACT

Oleo Resin Ginger	15 oz.
Oleo Resin Capsicum	2 oz.
Lemon Extract	5 pt.
Orange Extract	2½ pt.
Alcohol and Water	2 gal.

Use to:

1 gallon Simple Syrup.
3 ounces Extract.

GINGER ALE EXTRACT (Belfast)

Oleo Resin Ginger	24 oz.
Oleo Resin Capsicum	5½ oz.
Oil of Lemon (Terpeneless)	36 oz.
Oil of Orange (Terpeneless)	12 oz.
Oil of Cassia	1½ dr.
Oil of Rose, Artificial	½ dr.
Oil of Cloves	1½ dr.
Cologne Spirits	5½ gal.
Water	3 gal.

Use to:

1 gallon Simple Syrup.
2 ounces Extract.

GINGER ALE EXTRACT

Oil of Ginger	4 oz.
Oil of Capsicum	1 oz.
Lemon Extract	16 oz.
Orange Extract	8 oz.
Alcohol	3½ pt.
Water	3½ pt.

GINGER CHAMPAGNE SYRUP

Sugar	46 lb.
Water	6½ gal.
Soluble Extract of Ginger	8 fl. oz.
Soluble Extract of Orange	16 fl. oz.
Soluble Extract of Lemon	8 fl. oz.
Sulphurous Acid	4 fl. oz.
Vanilla Extract	2 fl. oz.
Sugar Color	2½ fl. oz.
Citric Acid Solution	32 fl. oz.

Use to:

⅛ ounce soda bottle.
1 to 1¼ ounces of above.

COLA FLAVOR

Oil of Lemon	120 drops
Oil of Sweet Orange	80 drops
Oil of Nutmeg	40 drops
Oil of Cinnamon	40 drops
Oil of Coriander	20 drops
Oil of Neroli	40 drops
Alcohol, 95 per cent	1 qt.

Add in rotation, shaking well before adding next ingredient, and let stand 48 hours, when it is ready to use.

COLA SYRUP

Sugar	60 lb.
Water	5 gal.

Dissolve sugar and bring to boil. Then, while boiling, stir in syrup:

Beet Sugar Color	40 fl. oz.
Let cool and add:	
Phosphoric Acid Syrup	3 fl. oz.
Alkaloid of Caffeine	1½ av. oz.
(Dissolved in 8 ozs. boiling water.)	
Fluid Extract Cola Leaves	1½ fl. oz.
Fluid Extract Kola Nuts	2½ fl. oz.
Alcohol	1 pt.
Extract of Vanilla	5 fl. oz.
Cola Flavor	4 fl. oz.
Glycerine	4 fl. oz.
Lime Juice	16 fl. oz.

Let age for three days.

Note: The sugar color used in all Cola drinks must be the best.

CONCRETE ORANGE EXTRACT

Gum Tragacanth	2 dr.
Glucose	4 oz.
Concentrated Oil of Orange	½ oz.
Oil of Bitter Orange	1 oz.
Concentrated Tincture of Orange	12 oz.
Citral	15 gr.

Use to:

1 gallon Simple Syrup.
1 ounce Extract.

ORANGE CHAMPAGNE SYRUP

Sugar	46 lb.
Water	6¼ gals.
Soluble Orange Extract	3 fl. oz.
Soluble Lemon Extract	20 fl. oz.
Soluble Lime Extract	5 fl. oz.
Sulphurous Acid	3 fl. oz.
Citric Acid Solution	3 fl. oz.
Orange color to suit.	

Use to:

Each pint bottle 2¼ ounces Extract.

DRY GINGER ALE EXTRACT

Solid Extract Jamaica Ginger	8 oz.
Oil Ginger	2 drams
Oil Sweet Orange	2 drams
Oil Limes, Distilled	1 dram
Oil Mace	¼ dram
Oil Coriander	¼ dram
Oil Lemon	¼ dram

Grind the above in a mortar with 4 ounces powdered magnesium carbonate; then add 1 gallon alcohol slowly while grinding in thoroughly; then add 1 gallon water slowly and stir thoroughly for 2 hours; add 2 ounces kieselguhr and filter through fine filter paper. The finished product should be aged to develop a finer aroma and taste.

Four ounces of this extract is used per gallon of syrup.

GINGER ALE

Jamaica Ginger, fine powder	8 lb.
Capsicum, fine powder	6 oz.
Alcohol, a sufficient quantity.	

Mix the powders intimately, moisten them with enough alcohol to make them distinctly damp but not wet, set aside for 4 hours, then pack in a cylindrical percolator and percolate with alcohol until 10 pints have been collected; place the percolate in a bottle of at least 2-gallon capacity and add 2 fluid drams of oleoresin ginger, shake and add 2½ pounds of finely powdered pumice stone and agitate frequently for 12 hours, then the next step is most important. Add 14 pints of water 1 pint at a time, then shake briskly and add the next, after adding all the water set aside for 24 hours, agitating strongly every hour or so, then add:

Oil of Lemon	1½ fl. oz.
Oil of Rose Geranium	3 fl. dr.
Oil of Bergamot	2 fl. dr.
Oil of Cinnamon	3 fl. dr.
Magnesium Carbonate	3 oz.

First rub the magnesia with the oils in a mortar, add 9 fluid ounces of the clear portion of the ginger mixture to which 2 ounces of alcohol have been added, and continue trituration, rinsing the mortar out with the ginger mixture. Pass the ginger mixture through a double filter and add the mixture of oils through the filter. Finally pass enough water through the filter to make 3 gallons of the finished extract, which is to be used 4 fluid ounces to a gallon of syrup. Dilute the syrup, 1 fluid ounce with 6 fluid ounces of carbonated water; bottle.

Note: The ginger ale can be colored a darker color with caramel.

SOLUBLE GINGER or CAPSICUM FLAVOR

12 Fluid Ounces Oleoresin Ginger or Capsicum.
243 Fluid Ounces Alcohol.
6 Ounces Precipitated Magnesium Carbonate.
189 Fluid Ounces Water.

In making the above flavor, first mix the oleoresin thoroughly with alcohol and then add the magnesium carbonate, working it into an even paste. Add the water slowly with thorough stirring, then filter.

SOLUBLE GINGER ALE EXTRACT

(To be used in the proportion of 4 ounces of extract to 1 gallon of syrup.)

Jamaica Ginger, in fine powder	8 lb.
Capsicum, in fine powder	6 oz.
Alcohol, a sufficient quantity.	

Mix the powders intimately, moisten them with a sufficient quantity of alcohol, and set aside for 4 hours. Pack in a cylindrical percolator and percolate with alcohol until 10 pints of percolate have resulted. Place the percolate in a bottle of the capacity of 16 pints, and add to it 2 fluid drams of oleoresin of ginger; shake, add 2½ pounds of finely powdered pumice stone, and agitate thoroughly at intervals of ½ hour for 12 hours. Then add 14 pints of water in quantities of 1 pint at each addition, shaking briskly meanwhile. This part of the operation is most important. Set the mixture aside for 24 hours, agitating it strongly every hour or so during that period. Then take

Oil of Lemon	1½ fl. oz.
Oil of Rose (or geranium)	3 fl. dr.
Oil of Bergamot	2 fl. dr.
Oil of Cinnamon	3 fl. dr.
Magnesium carbonate	3 fl. oz.

Rub the oils with the magnesia in a large mortar and add 9 ounces of the clear portion of the ginger mixture, to which have been previously added 2 ounces of alcohol, and continue trituration, rinsing out the mortar with the ginger mixture. Pass the ginger mixture through a double filter and add through the filter the mixture of oils and magnesia; finally pass enough water through the filter to make the resulting product measure 24 pints, or 3 gallons. If the operator should desire an extract of more or less pungency he may obtain his desired effect by increasing or decreasing the powdered capsicum in the formula.

LEMON EXTRACT (Terpeneless)

Oil of Lemon	30 lb.
Citral	8 oz.
Cologne Spirits	16 gal.

Put in a churn and work 2 hours. Of 11 gallons of water, add gradually about 5 gallons every hour and work for 2 hours more, then add 3 gallons water and work more. The whole process takes about 10 hours. After 10 hours add 1½ gallons Cologne Spirits. Let stand for 48 hours and filter.

Use to:

1 gallon Simple Syrup.
1 ounce Extract.

ROOT BEER

A

Oil of Sweet Birch or Methyl Salicylate	15	oz.
Oil of Cloves	¾	oz.
Oil of Sassafras	⅜	oz.
Oil of Lemon	½	oz.
Oil of Cassia	⅛	oz.
Mexican Vanilla Extract (best quality)	6	pt.
Cologne Spirits	15	pt.
Caramel	4	oz.
Oil of Nutmeg	2	dr.

Dissolve the oils in the alcohol. Mix the caramel with the vanilla extract. Pour the colored vanilla extract into the alcoholic solution. A brown precipitate will form, which acts as the clarifying agent, and may be filtered out after standing an hour or 2 hours.

Use to:

1 gallon Simple Syrup (10 pounds sugar to gallon water).
1 ounce Root Beer Extract.
¼ ounce Citric Acid Solution.
2 ounces Caramel.

B

Oil of Wintergreen	20	oz.
Oil of Sassafras	24	oz.
Oil of Anise	10	dr.
Oil of Cassia	1	oz.
Cologne Spirits	3½	gal.
Water	½	gal.

Use to:

1 gallon Simple Syrup.
2 ounces Extract.

C

Oil of Sassafras	2½ fl. oz.
Oil of Wintergreen	2½ fl. oz.
Oil of Sweet Orange	2 fl. oz.
Amyl Butyrate	2 fl. oz.
Oil of Spruce	½ fl. oz.
Oil of Cloves	2 dr.
Oil of Anise	2 dr.
Cologne Spirits	7 pt.
Water	2 pt.

Add a little at a time the oils to the Cologne Spirits, shake well, add the 2 pints of water, and filter through pumice.

Use to:

1 gallon Simple Syrup.
1½ ounces Extract.

D

Oil of Sassafras	5 fl. oz.
Oil of Peppermint	½ fl. oz.
Oil of Tar	10 drops
Oil of Cinnamon	10 drops
Carbonate Magnesium	4 av. oz.
Cologne Spirits	½ gal.
Water	½ gal.

Use to:

1 gallon Simple Syrup.
½ fl. ounce Extract.
2 fl. ounces Sugar Color.

SARSAPARILLA EXTRACT

Oil of Wintergreen	4 oz.
Oil of Sassafras	4 oz.
Oil of Anise	1 oz.
Cologne Spirits	5 pt.
Powdered Pumice Stone	4 oz.
Granulated Sugar	8 oz.
Water	2½ pt.
Sugar Color	1 oz.

Dissolve the oils in 2 pints of the spirits. Each oil must be added separately and well shaken with the spirits before another oil is added. Now put the pumice stone and sugar in a Wedgewood mortar, add the mixture gradually, and rub together to a paste. Mix the remainder of the spirits and water together, add the sugar color to these, and dissolve carefully. Mix the whole together gradually, stirring well until all combines, and filter through filter paper.

Use to:

1 gallon Simple Syrup.
1 ounce Extract.

Cook mixture of cocoa and water in double boiler for 5 minutes. Add sugar, stir until dissolved, and then add milk which has been previously preheated to 100° F. Add vanilla extract and heat to 120° F. Stir in junket tablets which are dissolved first in a little water. Pour into containers immediately, let stand until set.

FOOD COLORS

The following food colors are permitted by the United States Department of Agriculture for use in food products. It is, however, necessary to state their presence on the label as

ARTIFICIALLY COLORED or

CERTIFIED FOOD COLORS ADDED

PRIMARY COLORS

RED	Erythrosine	
	Amaranth	
	Ponceau 3R	
	Ponceau SX	
ORANGE	Orange I	
GREEN	Guinea Green	
	Light Green S. F.	
	F. C. F. Green	
BLUE	Sodium Indigo Disulphonate	
YELLOW	Tartrazine	
	Yellow A. B.	Soluble in oil only
	Yellow O. B.	Soluble in oil only

SECONDARY AND TERTIARY COLORS

Are mixtures of two and three or more primary colors.

SOLUBILITY IN WATER

Color	Maximum Solubility Ounces Per Gallon
Amaranth	13.3
Orange I	9.3
Light Green S. F.	22.2
Guinea Green	17.6
Ponceau 3R	30.0
Sodium Indigo Disulphonate	2.7
Tartrazine	16.7
Erythrosine	13.0

SOLUBILITY IN ALCOHOL

Color	Solubility in 95% Alcohol	Solubility in 50% Alcohol
Amaranth	Insoluble	Less than ½ oz. per gal.
Ponceau 3R and Ponceau SX	Insoluble	As good as water
Sodium Indigo Disulphonate	Insoluble	Less than ½ oz. per gal.
Tartrazine	Insoluble	As good as water
Light Green S. F.	Slightly	Easily in 70% alcohol
Guinea Green	Fair	Better in 85% alcohol
Erythrosine	Fair	Better in 85% alcohol
Orange I	Fair	Better in 85% alcohol

For making solutions in diluted alcohol, dissolve the color in the water first, before incorporating with the alcohol required.

COMPLEMENTARY COLORS

Blue + Red + Yellow → Gray
Red + Green → Gray
Blue + Orange → Gray
Yellow + Violet → Gray

SECONDARY FOOD COLORS

Mixtures Compounded from Primary Colors

COLOR SCHEME

YELLOWS
Yolco Yellow
Egg Color
Canary Yellow
Lemon Yellow
Golden Yellow
Brilliant Yellow
Macaroni Yellow
Ice Cream Color
Ice Cream Cone Color
Cheese Color

ORANGES
Blood Orange
Brilliant Orange
Yellow Orange
Yellowish Orange
Red Orange
Reddish Orange
Peach

VIOLETS
Violet
Royal Purple
Lavender
Grape

GREENS
Dark Green
Light Green
Yellow Green
Blue Green
Myrrteline (Olive Green)
Mint Green
Creme de Menthe Green
Leaf Green

BLUES
Dark Blue
Light Blue
Auramine Royal Blue
Sky Blue

REDS
Deep Red
Ruby Red
Striping Red
Striping Red (Peppermint Sticks)
Carmine Red
Cherry Red
Sherry Red
Raspberry Red
Strawberry Red
Grenadine
Imperial Rose
Brilliant Red
Pink
Persian Rose
Brilliant Rose
Rose (Bluish)
Rose (Pink)

BROWNS
Brown
Almond Brown
Reddish Brown
Chocolate Brown
Burnt Peanut Brown
Loganberry

SECONDARY FOOD COLORS

YELLOWS		
YOLCO YELLOW (for bakers)		
	Tartrazine	9 oz.
	Amaranth	1 oz.
	or	
	Orange I	
EGG SHADE		
	Tartrazine	14½ oz.
	Orange I	1½ oz.
CANARY YELLOW, LEMON YELLOW (for confectioners)		
	Straight Tartrazine	
GOLDEN YELLOW		
	Tartrazine	75 oz.
	Orange I	25 oz.
	or	
	Tartrazine	15¼ oz.
	Orange I	¾ oz.
BRILLIANT YELLOW (macaroni)		
	Tartrazine	
ORANGES		
PEACH COLOR		
	Ponceau	1 lb. 4 oz.
	Tartrazine	10 oz.
YELLOW ORANGE		
	Orange I	9 oz.
	Tartrazine	1 oz.
ORANGE (Yellowish)		
	Tartrazine	50 oz.
	Orange I	50 oz.
REDDISH ORANGE		
	Orange I	9 oz.
	Amaranth	1 oz.
ORANGE (Reddish)		
	Ponceau	75 oz.
	Tartrazine	30 oz.
BRILLIANT ORANGE		
	Orange I	
VIOLETS		
VIOLET		
	Amaranth	88 oz.
	Sodium Indigo Disulphonate	42 oz.
	or	
	Amaranth	8 oz.
	Sodium Indigo Disulphonate	8 oz.
ROYAL PURPLE		
	Amaranth	8 oz.
	Sodium Indigo Disulphonate	2 oz.

GRAPE		
	Amaranth	2 oz.
	Sodium Indigo Disulphonate	¼ oz.
	or	
	Amaranth	
	Sodium Indigo Disulphonate	4 oz.
GREENS		
GREEN		
	Tartrazine	6 or 6.5 oz.
	Sodium Indigo Disulphonate	4 or 3.5 oz.
BRILLIANT GREEN		
	Tartrazine	10 oz.
	Sodium Indigo Disulphonate	6 oz.
YELLOW GREEN		
	Sodium Indigo Disulphonate	4 oz.
	Tartrazine	12 oz.
MINT GREEN		
	Tartrazine	7 oz.
	Sodium Indigo Disulphonate	3 oz.
	or	
	Tartrazine	8 oz.
	Sodium Indigo Disulphonate	8 oz.
CREME de MENTHE GREEN		
	Light Green S. F.	48 oz.
	Tartrazine	52 oz.
MYRRTELINE (Olive Green)		
	Sodium Indigo Disulphonate	4 oz.
	Orange I	1½ oz.
	Tartrazine	10½ oz.
LEAF GREEN		
	Light Green S. F.	
REDS		
DEEP RED		
	Ponceau	75 oz.
	Amaranth	25 oz.
RUBY RED		
	Amaranth	
STRIPING RED		
	Erythrosine	10 oz.
	Amaranth	30 oz.
	Ponceau	60 oz.
STRIPING RED (Peppermint Sticks)		
	Orange I	2 oz.
	Amaranth	14 oz.
CARMINE RED		
	Erythrosine	90 oz.
	Ponceau	10 oz.
	or	
	Tartrazine	4 oz.
	Amaranth	12 oz.

CHERRY RED		
	Ponceau	
	or	
	Amaranth	68 oz.
	Orange I	32 oz.
SHERRY RED		
	Amaranth	12 oz.
	Orange I	4 oz.
RASPBERRY RED		
	Amaranth	
GRENADINE RED		
	Amaranth	
STRAWBERRY RED		
	Amaranth	8.5 oz.
	Orange I	1.5 oz.
	or	
	Amaranth	94 oz.
	Ponceau	16 oz.
	or	
	Amaranth	30 oz.
	Ponceau	70 oz.
	or	
	Amaranth	13 oz.
	Orange I	3 oz.
BRILLIANT RED (Imperial Rose)		
	Amaranth	14 oz.
	Orange I	2 oz.
PERSIAN ROSE		
	Erythrosine	88 oz.
	Tartrazine	8 oz.
	Amaranth	4 oz.
BRILLIANT ROSE		
	Amaranth	14 oz.
	Orange I	2 oz.
ROSE (Pink)		
	Amaranth	12 oz.
	Sugar	4 oz.
ROSE (Bluish)		
	Erythrosine	50 oz.
	Amaranth	5 oz.
PINK		
	Amaranth	15½ oz.
	Sodium Indigo Disulphonate	½ oz.
BROWNS		
BROWN		
	Orange I	13 oz.
	Sodium Indigo Disulphonate	8 oz.
	Amaranth	3 oz.
	Tartrazine	2 oz.

ALMOND BROWN (Reddish brown)		
	Orange I	13¾ oz.
	Amaranth	2¼ oz.
CHOCOLATE BROWN		
	Sodium Indigo Disulphonate	28 oz.
	Amaranth	20 oz.
	Tartrazine	40 oz.
	Orange I	12 oz.
	or	
	Sodium Indigo Disulphonate	4 oz.
	Amaranth	4 oz.
	Tartrazine	6 oz.
	Orange I	6 oz.
BURNT PEANUT BROWN		
	Orange I	45 oz.
	Ponceau	25 oz.
	Light Green S. F.	30 oz.
BLUES		
AURAMINE ROYAL BLUE		
	Sodium Indigo Disulphonate	12 oz.
	Sugar	4 oz.

TERTIARY COLORS

Mixtures from Secondary Colors

ICE CREAM COLOR		
	Egg Color Yellow	4 oz.
	Strawberry Red	½ oz.
ICE CREAM CONE COLOR		
	Tartrazine	7 oz.
	Chocolate	1 oz.
BLOOD ORANGE		
	Egg Color	6¾ oz.
	Strawberry Red	1¼ oz.
CHEESE COLOR		
	Ice Cream Color	½ oz.
	Water	½ oz.

The so-called "Buttercup Yellow" is Zinc Chromate ($ZnCrO_4$), which is prohibited by law for use in any food preparation.

POWDERED COLORS

READY TO USE COLOR MIXTURES

General Formula

Primary or Secondary Color	6 oz.
Salt or Sugar	10 oz.

YELLOWS

EGG COLOR

I. Tartrazine	9	lb.
Orange I	2	lb.
Salt	36	lb.
II. Tartrazine	5	lb.
Orange I	14	oz.
Sugar	13	lb.
III. Tartrazine	16	lb.
Orange I	20	oz.
Sugar	34	lb.
IV. Tartrazine	18	oz.
Orange I	1	oz.
Sugar	30	oz.
V. Tartrazine	4	oz.
Orange I	½	oz.
From No. V.	2½	oz.
Salt or Sugar	13½	oz.

LEMON COLOR

Tartrazine	6 oz.
Salt or Sugar	10 oz.

PEACH COLOR

Secondary Peach Color	6 oz.
Salt or Sugar	10 oz.

BROWNS

Brown
Chocolate Brown
Burnt Peanut Brown

Take from these secondary colors	6 oz.
Salt or Sugar	10 oz.

GREENS

Dark Green
Light Green
Green
Mint Green
Creme de Menthe Green
Leaf Green

For Dark Green (bluish) take Guinea Green
For Light Green (yellowish) take Light Green S. F.
For the other shades:

Secondary Colors	6 oz.
Salt or Sugar	10 oz.

ORANGES

Yellow Orange
Reddish Orange

Take from these secondary colors	6 oz.
Salt or Sugar	10 oz.

VIOLETS

Violet
Royal Purple
Grape

Take from these secondary colors	6 oz.
Salt or Sugar	10 oz.

RED

Ded Red
Striping Red
Carmine Red
Cherry Red
Raspberry Red
Strawberry Red
Persian Rose
Rose (Bluish)
Brilliant Rose

Take from these secondary colors	6 oz.
Salt or Sugar	10 oz.

BLUES

Take Sodium Indigo Disulphonate	6 oz.
Salt or Sugar	10 oz.

NOTE: When using salt a very fine grade should be employed.

When using sugar, moisten with a little color dissolved in water, let dry, and grind through a very fine sieve to a fine powder. Then mix with the remaining part of the color. In this way deep shades of color mixtures are obtained.

Small quantities of color mixtures may be mixed by hand and sifted through a fine-mesh sieve. For larger quantities a mixing machine, as used for beating egg whites and meringues, can be used to great advantage. In this case the mixing bowl should be provided with a cover to prevent the color from flying around.

LIQUID COLORS

GENERAL FORMULA

Primary or Secondary Color	4	oz.
Glycerin	6	oz.
Balance Water for	1	gal.

YELLOWS

EGG COLOR (A very popular formula)

Tartrazine	5	lb. 6 oz.
Orange I	1	lb. 4 oz.
Glycerin	1¼	gal.
Balance Water for	50	gal.

(That is approximately 2.1 ounces of color per gallon.)

Or

Tartrazine	½	lb.
Orange I	¾	oz.
Alcohol	4	oz.
Balance Water for	2	gal.

LEMON

Tartrazine	25	oz.
Glycerin	30	oz.
Balance Water for	5	gal.

BROWN

Primary or Secondary Color	4	oz.
Glycerin	6	oz.
Balance Water for	1	gal.

GREENS

Guinea Green	12	oz.
Tartrazine	12	oz.
Glycerin	18	oz.
Balance Water for	6	gal.

ORANGES

Orange I	2½	lb.
Alcohol	1	qt.
Balance Water for	5	gal.

Boil and strain.

VIOLETS

Amaranth	2	oz.
Sodium Indigo Disulphonate	2	oz.
Glycerin	6	oz.
Balance Water for	1	gal.

LAVENDER

Erythrosine	2	oz.
Sodium Indigo Disulphonate	2	oz.
Balance Water for	1	gal.

GRAPE

Amaranth	3	oz.
Sodium Indigo Disulphonate	1	oz.
Glycerin	6	oz.
Balance Water for	1	gal.

REDS

Amaranth	2½	lb.
Alcohol	½	gal.
Balance Water for	10	gal.

RASPBERRY

As Red Color

STRAWBERRY

Amaranth	17	oz.
Ponceau	3	oz.
Balance Water for	5	gal.

CHERRY		
Ponceau	4	oz.
Alcohol	6	oz.
Balance Water for	1	gal.
PINK		
Erythrosine	4	oz.
Glycerin	6	oz.
Balance Water for	1	gal.
BRILLIANT ROSE		
Amaranth	3½	oz.
Orange I	½	oz.
Glycerin	6	oz.
Balance Water for	1	gal.
BLUES		
Sodium Indigo Disulphonate	4	oz.
Glycerin	6	oz.
Balance Water for	1	gal.
BUTTER COLOR		
Yellow O. B.		
Or		
Yellow A. B.	2½	oz.
Cottonseed Oil	1	gal.

NOTE: Dissolve the colors in hot water and strain through cheesecloth. Where alcohol is required, add the alcohol after cooling. Alcohol and glycerin act as a preservative. The addition of 1 to 2 ounces of citric or tartaric acid is also said to have preserving effects.

PASTE COLORS

GENERAL FORMULA		
Color	1	lb.
XXXX Sugar	8	lb.
Glycerin	3½	pt.

Heat glycerin to about 145° F. Add the color. Stir until all is dissolved. Then add gradually, under good stirring, 8 pounds XXXX sugar (Confectioner's), which has been previously sifted. Continue stirring on low gas until a smooth paste is obtained. Fill while still hot.

YELLOW		
Tartrazine	13½	oz.
Orange I	1½	oz.

Proceed as above.

Because of the partly insoluble tartrazine, strain the paste.

LEMON		
Tartrazine	1	lb.
BROWN		
Brown (Secondary Color)	1	lb.

GREEN	
Guinea Green	1 lb.
ORANGE	
Orange I	1 lb.
VIOLET	
Sodium Indigo Disulphonate	3 oz.
Amaranth	12 oz.
RED	
Amaranth	1 lb.
PINK	
Erythrosine	1 lb.
BLUE	
Sodium Indigo Disulphonate	1 lb.
BLACK	
Black (Wood Charcoal)	3 oz.
Grape Sugar	1 oz.
Water	6 oz.

VEGETABLE COLORS

GREENS	1. CHLOROPHYLL
	2. By mixing TINCTURE OF SAFFRON or TINCTURE OF TURMERIC with a solution of INDIGO CARMINE (made from the paste) in various proportions a variety of green shades can be obtained.
REDS	COCHINEAL SYRUP and CARMINE
YELLOWS	TINCTURE OF SAFFRON and TINCTURE OF TURMERIC
BROWNS	CARAMEL (Burnt Sugar)
BLUES	INDIGO CARMINE PASTE
VIOLET	By mixing RED + BLUE
ORANGES	TINCTURE OF ANNATTO and by mixing RED + YELLOW
GRAYS	By mixing RED + GREEN or RED + BLUE + YELLOW

TINCTURE OF CHLOROPHYLL

Extract green spinach with diluted alcohol and evaporate to paste. Use 1 part paste to 10 parts alcohol and filter—or—moisten fine, hacked grass with water and press. Boil the juice and decant the greenish liquid. Press the residue off from the surplus liquid and extract the residue with alcohol. Filter and evaporate. Take up the residue in hot water. The insoluble separating greenish flocky mass is dissolved in hydrochloric acid and finally precipitated with water.

TINCTURE OF SAFFRON

Saffron	3 oz.
Water	1 qt.
Alcohol	1 qt.

Macerate several days and filter.

TINCTURE OF TURMERIC

Turmeric (Bruised)	1 oz.
Diluted Alcohol (50–50)	6 oz.

Macerate in a closed vessel for 7 days and filter.

COCHINEAL SYRUP

Powdered Cochineal	12 oz.
Potassium Bicarbonate	4 oz.
Distilled Water	30 oz.
Alcohol	24 oz.

Rub the potassium bicarbonate with the cochineal powder. Mix the alcohol and water and add to the powder. Filter, then mix the solution with

Syrup	120 oz.

thoroughly.

TINCTURE OF CARMINE

Carmine	22 oz.
Strong Ammonia Water (Enough to dissolve the carmine)	
Distilled Water, enough to make	500 oz.

Dissolve the carmine in the ammonia water and add the distilled water.

EFFECT OF METALS ON CERTIFIED FOOD COLORS

Copper		Aluminum	
Plain Water	Acid Solution	Plain Water	Acid Solution
Ponceau 3 R			
Solution clear. Pour-out on paper equal in strength and bluer in shade.	Solution duller and cloudy. Pour-out on paper equal in strength and bluer in shade.	Solution clear. Pour-out on paper about equal in strength and shade.	Solution clear. Pour-out on paper slightly weaker, slightly bluer.
Amaranth			
Solution cloudy, dark brown.	Solution cloudy, dark brown.	Solution clear, slightly yellower.	Solution clear, much weaker.
Orange I			
Solution cloudy. Pour-out on paper very much redder and slightly weaker.	Solution brown and muddy. Pour-out on paper, color completely destroyed.	Solution clear, slightly redder, equal in strength.	Solution dark and muddy. Pour-out on paper, color entirely destroyed.
Tartrazine			
Solution clear, slightly redder, equal in strength to standard. Pour-out on paper equal in strength to standard.	Solution very red and clear. Pour-out on paper shows color redder and 50% destroyed.	Solution clear, equal in strength and little, if any, change in shade. Same pour-out on paper.	Little change in shade. Solution clear, color destroyed to some extent, about 75% strength of standard. Pour-out on paper indicates little change in shade, about 50/75% strength.
Light Green S F Yellowish			
Little change.	Color practically destroyed. Almost clear.	Little change.	Color practically destroyed. Almost clear.
Fast Green F C F			
Little change.	Solution brownish.	Little change.	Solution little weaker.
Erythrosine			
Copper has no effect on dye in city water. Same applies to acid solution.		Aluminum precipitates dye somewhat although it does not affect shade when poured on paper. This applies to both plain water and acid solution.	

Copper		Aluminum	
Plain Water	Acid Solution	Plain Water	Acid Solution
Naphthol Yellow S Clarity of solution is unchanged but shade turns appreciably redder.	Same as for plain water.	Shade becomes slightly redder and clarity of solution is unchanged.	Same as for plain water.
Ponceau S X Solution turns very dark and shade becomes considerably yellower and duller.	Same as for plain water.	No apparent change affected.	No apparent change affected.
Sunset Yellow F C F Solution becomes opaque and very dark brown. Shade turns much duller.	Same as for plain water.	Clarity of solution unchanged. Color becomes slightly redder.	Same as for plain water.
Guinea Green B No changes noted but an oily film appears on surface of liquid.	Same as for plain water.	Unaffected.	Unaffected.
Brilliant Blue F C F Unaffected.	Unaffected.	Unaffected.	Unaffected.
Sodium Indigo Disulfonate Solution unchanged, slight loss of strength.	Practically unaffected.	Solution unchanged, slight loss of strength.	Solution unchanged, appreciable loss in strength.

NOTE: Our observation on this color is that it appears to be readily reducible, affecting the strength, whereas the color remaining unreduced seems to be unaffected and further reduction will tend to reduce it to a colorless compound.

TINCTURE OF INDIGO CARMINE

Dissolve in distilled water any amount of the indigo carmine paste to get the desired strength or shade. Usually

Paste	1 oz.
Warm Water	20 oz.

Dissolve, filter, and add 5% alcohol.

TINCTURE OF ANNATTO

One hundred ounces of annatto are extracted with 150 ounces of water, containing 1½% potassium carbonate. Evaporating to 60 ounces, add 12 ounces of alcohol and filter.

Used for coloring butter and cheese.

SUGAR COLORING (Caramel Color)

Add 2 gallons of water to 10 gallons of caramel.

COLORING MATERIAL FOR LIQUEURS

RED
1. Steep 4 ounces of raspings of RED SANDERS WOOD in a pint of strong spirits of wine for 2 weeks. Strain and filter.
2. Steep 3 ounces of COCHINEAL, finely powdered, in a pint of strong spirits of wine for 2 weeks; add 2 drams of powdered alum and filter clear.

BLUE
Steep 4 drams of INDIGO in a bottle with 2 ounces of SULPHURIC ACID for several days, frequently putting the bottle into hot water, add ½ pint (8 ounces) of distilled water and filter.

YELLOW
Steep 1 ounce of SAFFRON in ½ pint of spirits of wine for a week; filter.

GREEN
Mix equal parts of Blue and Yellow coloring.

VIOLET
Mix 1 part of Blue with 2 of Red coloring.

PINK
Steep 4 ounces of CUDBEAR in a quart of strong spirits for 2 weeks; filter.

BROWN
Add 2 gallons of water to 10 gallons of CARAMEL.

BEVERAGES

The essential principles are extracted by

INFUSION with hot spirit (if the aromatic principle is not too volatile), the ingredients being first moistened with boiling water; or by

MACERATION with cold spirits, which may require hours, days, or even months, according to the solubility of the essential principles; by DIGESTION or prolonged infusion or maceration in hermetically closed vessels, when the aromatic principles are extracted only with difficulty by cold alcohol; or by

DISTILLATION, when only the natural essential oils or other volatile flavoring constituents are required.

The following plants should always be distilled by steam: Absinth, Anise, Balms, Caraway, Citronella, Fennel, Juniper, Hyssop, Lavender, Melilot, Mint, Orange, Rose, Sage, Thyme.

Aromatic tinctures are prepared by filtration or percolation.

PERCOLATION

Moisten the powdered drug or mixed drugs with a sufficient quantity of the menstruum to rend it evenly and distinctly damp. Transfer it to a percolator and, without pressing the powder, allow it to stand well covered for 6 hours. Then pack it firmly (unless otherwise stated) and pour on enough of the menstruum to saturate the powder and leave a menstruum above it. Close when dripping begins, cover the percolator, macerate for 24 hours. Let drip slowly.

CLARIFICATION OR COLLAGE

with fining materials is best avoided if possible but is often necessary. The principal materials and the proportions used are as follows:

Albumen (White of Egg)
  1 egg for every 10 litres
Gelatin
  30 grams per 100 litres
Isinglass
  10 grams per 100 litres
Milk
  1 litre boiled in 100 litres with 15 grams Alum
Animal or Vegetable Carbon
Alum and various "filtering powders" are also employed
For colored liquids filtration is preferable to collage
Tannin and Gelatin Combinations
Bentonite

BITTERS

They are prepared by maceration without distillation and are considered to be tonic and stomachic and to improve the appetite when taken in moderation.

A bitter made according to the following formula, which represents the proportions necessary to make 20 gallons, may be taken as typical:

Gentian	6 lb.
Cinnamon	5 lb.
Caraway	5 lb.
Juniper Berries	1 lb.
Cloves	½ lb.

Macerate in 7 gallons of 60% alcohol for 10–12 days. Strain, filter, add syrup consisting of 10 pounds of sugar to 13 gallons of water and color with cochineal.

ENGLISH BITTERS

Digest

Outer Rind of Dried Orange Peel	5	oz.
Cinchona Bark	6	oz.
Gentian	6	oz.
Carduus Benedictus	8	oz.
Centaury	8	oz.
Wormwood	8	oz.
Orris Root	4	oz.

with 12 gallons of proof spirits and 12 pounds of sugar.

ORANGE BITTERS

Orange Peel, Ground	5	lb.
Coriander Seed, Ground	2	lb.
Lemon Peel, Ground	1	lb.
Saccharin (1 : 300)	¼	oz.
Alcohol	8	gal.
Distilled Water	12	gal.

Macerate 14 days with a portion of the alcohol and water (enough to cover the drugs). Percolate with the remainder. Filter, if necessary.

ANGOSTURA TYPE BITTERS

Gentian Root	4	oz.
Calisaya Bark	10	oz.
Canada Snake Root	10	oz.
Virginia Snake Root	10	oz.
Licorice Root	10	oz.
Yellow Bark	10	oz.
Allspice	10	oz.
Dandelion Root	10	oz.
Angostura Bark	10	oz.
Cardamom Seeds	6	oz.
Tolu Balsam	4	oz.
Galanga	4	oz.
Rhubarb	4	oz.
Orange Peel	1	lb.
Alkanet Root	1	lb.
Caraway Seed	1½	oz.
Cinnamon	1½	oz.
Cloves	½	oz.
Catechu	2	oz.
Coriander Seed	2	oz.
Nutmeg	2	oz.
Wormwood	2	oz.
Mace	1	oz.
Red Sanders Wood	1¼	lb.
Turmeric	8	oz.

Macerate 15 days in 50 gallons of proof spirits. Add 30 pounds of honey and filter.

CHAPTER V

GELATIN DESSERTS AND PUDDING POWDERS

The popularity of gelatin desserts has increased considerably for a number of reasons, namely, the overcoming of lumpiness in the product, the marked improvement in flavor, and, last, the production of a neutral-smelling and tasting gelatin, which, of course, is the base of these products.

A poor gelatin, high in bacteria, weak in jell strength, plus an inferior quality of fruit flavor, will never build up a repeat business. Unless a quality product is made, it should not be made at all. To make a free-flowing gelatin dessert, use some dextrose as a base for compounding the flavor. It is the dextrose that prevents the lumping.

In the manufacturing of pudding powders, use the best quality of starch and cocoa powders. All these products should be sifted before final packaging.

GELATIN DESSERTS

INSTANT JELL

XXXX Sugar	14 lb.
Granulated Sugar	17½ lb.
Gelatin	4½ lb.
Tartaric Acid	11 oz.
Flavor	8 oz.
Color	To suit

Directions: Dissolve 1 pound in 5 pints of boiling water. Stir until entirely dissolved.

This preparation is especially made for restaurants, institutions, etc. Easy and very economical to prepare.

JELLY POWDERS
(For Household Use)

STOCK MIXTURE

Gelatin	14 lb.
Sugar	85 lb.
Corn Sugar	15 lb.
Tartaric Acid	1 lb. 3 oz.
Cream of Tartar	2 lb. 6 oz.

To this mixture add the coloring and flavoring mixture of the flavor wanted.

Directions for use: Dissolve 3¼ ounces of the jelling powder (92 grams) in ½ pint of boiling water, stir until dissolved. Add ½ pint of cold water. Stir, pour into mold and chill. If fruits, nuts, etc., are to be used, add them after the gelatin dessert is cold and set sufficiently to hold them in place. All fruits, with the exception of pineapple, may be used either raw or cooked. Pineapple must be cooked first or the dessert will not set. In cup measure: ½ cup of powder, 1 cup of hot water, add 1 cup of cold water, proceed as above.

COLOR SOLUTIONS

Amaranth	13.3 oz.	1 gal. Water
Orange I	6.5 oz.	1 gal. Water
Tartrazin	13.0 oz.	1 gal. Water
Green	13.0 oz.	1 gal. Water

Dissolve the color in hot water.

PREPARATION OF COLORING AND FLAVORING MIXTURE FOR JELLY POWDER MANUFACTURE

Mixture to be added to a 118-pound batch:

STRAWBERRY	
Strawberry Extract, True Fruit	174 cc.
Strawberry Flavor, Imitation	4½ cc.
Amaranth Solution	42 cc.
Orange I Solution	21 cc.
RASPBERRY	
Raspberry Extract, True Fruit	174 cc.
Raspberry Flavor, Imitation	9 cc.
Amaranth Solution	68 cc.
CHERRY	
Cherry Extract, True Fruit	174 cc.
Cherry Flavor, Imitation	2 cc.
Cherry Extract, Imitation	35 cc.
Amaranth Solution	55 cc.
PINEAPPLE	
Pineapple Extract, True Fruit	174 cc.
Pineapple Flavor	35 cc.
Tartrazin Solution	43 cc.
LEMON	
Lemon Oil, Concentrated	52 cc.
Tartrazin Solution	32 cc.
ORANGE	
Orange Oil, Concentrated	52 cc.
Orange Solution	27 cc.
LIME	
Lime Oil, Concentrated	52 cc.
Tartrazin Solution	7 cc.
Light Green S.F. Solution	32 cc.
MINT	
Terpeneless Spearmint Oil	52 cc.
Tartrazin Solution	19 cc.
Light Green S.F. Solution	24 cc.

PUDDING POWDERS

CHOCOLATE PUDDING

	Formula No. 1
Sugar	120 lb.
Cocoa	30 lb.
Cornstarch	50 lb.
Salt	2 lb.
Vanillin	3 oz.

CHOCOLATE PUDDING

	No. 2
Sugar	120 lb.
Cornstarch	50 lb.
Tapioca Starch	15 lb.
Cocoa	20 lb.
Ground Chocolate	20 lb.
Salt	2 lb.
Vanillin	1 oz.

Directions for use (4-oz. package): Mix thoroughly contents of this package with ½ cup of milk. Add 1½ cups of milk and place over fire, stirring constantly until it thickens and reaches boiling point. Remove from fire, pour into mold or cups, which have been dipped in cold water. Cool. Serve cold, either plain or with whipped cream. Serves four.

VANILLA PUDDING

	Formula No. 1
Sugar	64 lb.
Cornstarch	26¾ lb.
Milk Powder	7½ lb.
Egg Yolk, Powdered	1¼ lb.
Salt	4¾ oz.
Vanillin	1⅛ oz.
Glycerin	10 fl. oz.
Color (Powder egg shade)	0.15 oz.

Directions for use: To 4 ounces of vanilla pudding add ½ cup of cold water, and stir mixture to a smooth paste. Bring 1½ cups of water to a boil and add the vanilla pudding mixture, stirring constantly to prevent scorching. Continue stirring until dessert comes to a thorough boil. Then, while still hot, pour into molds or sherbet glasses and allow to cool in the ice box. If desired use milk instead of water.

VANILLA PUDDING

	No. 2
Cornstarch	18 lb.
Sugar	60 lb.
Corn Sugar	15 lb.
Salt	8 oz.
Vanillin Flavor	To suit

Directions: Use 4 ounces per pint of milk.

Use the formulae as given under "Pie Filling Powders" for LEMON PUDDING, CARAMEL PUDDING, CHERRY PUDDING, BUTTERSCOTCH PUDDING, ORANGE PUDDING, STRAWBERRY PUDDING, RASPBERRY PUDDING.

If sugar is to be added to the mixture, add sugar to about 2½ times the weight of the mixture. The directions can then be followed as given under vanilla pudding.

CUP CUSTARD

Starch	1 lb.
Gum Tragacanth	2 oz.
Turmeric	150 gr.
Bitter Almond Oil	30 min.
Lemon Essence	1 dr.

Put up in 1-ounce packages for 1 pint of milk.

Directions: One package to be stirred in a dish with 2 tablespoonsful of milk, boil the remainder of the milk with 2 ounces of sugar, and while boiling pour gradually into the dish, stirring briskly. Bake as a custard.

CUP CUSTARD

Cornstarch	1 lb.
Arrowroot	1 lb.
Turmeric	2 dr.
Almond Oil	10 drops
Nutmeg Oil	3 drops

Use above directions.

Or

Cornstarch	1 lb.
Turmeric	1 dr.
Bitter Almond Oil	10 drops

Use above directions.

CUP CUSTARD

Whole Egg Powder	2½ lb.
Whole Milk Powder	4 lb.
XXXX Sugar	3 lb.
Salt	¼ oz.
Vanilla Extract	¼ tsp.

Use water only in preparing.

Dissolve a quantity of the powder in enough cold water to make a thin cream and cook in a double boiler, stirring while cooking, and when thick enough pour into molds.

FRENCH CUSTARD CREAM

Whole Milk Powder	9 oz.
Granulated Sugar	24 oz.
Cornstarch	5 oz.
Powdered Egg Yolk	2 oz.
Shortening	2 oz.
Salt	¼ oz.
Vanillin	.10 g.

Mix and sift the above ingredients. Whip into a smooth paste with 1 pint of water, and then add to 3 pints of boiling water, and stir constantly with a wire whip until the mixture thickens. Set aside to cool.

CUSTARD PIE FILLING

	Formula No. 1
Whole Milk Powder	3 oz.
Granulated Sugar	24 oz.
Salt	1 oz.
Cornstarch	3 oz.
Vanillin	.1 g.
Powdered Whole Eggs	5 oz.

Dissolve the above in 1 pint of water and then pour this into 8 pints of boiling water and stir until it thickens.

CUSTARD PIE FILLING

	Formula No. 2
Powdered Skimmed Milk	4 oz.
Granulated Sugar	5 oz.
Shortening	1 oz.
Powdered Whole Eggs	1 oz.
Tapioca Starch	1½ oz.
Vanilla and Mace for flavoring	

Beat into a smooth paste with 8 ounces of water, and then add to 1½ pints of boiling water, and heat until it thickens.

COCOANUT PUDDING

XXXX Sugar	1 lb.
Cornstarch	1 lb.
Powdered Milk	4 oz.
Gelatin	¼ oz.
Cocoanut	6 oz.
Vanilla Flavor	

Directions: Use as cup custard (water only).

COCOANUT CREAM CUSTARD

Cornstarch	1 lb.
Gelatin	1 dr.
Granulated Cocoanut	2 oz.
Vanillin	1 dr.
Turmeric	1 dr.

Directions: Mix with cold milk to make creamy. Dissolve ½ cup of sugar in 2 cups of hot milk. Pour creamed filling into the hot milk and stir on slow fire until it thickens. Continue to stir for 3 minutes after it thickens. Remove from fire and beat thoroughly until cool. Serve in sherbet glasses with whipped cream.

PUDDING POWDERS (German Recipes)

BASE

Starch	500 g.
Powdered Sugar	500 g.
Bicarbonate of Soda	20 g.
Potassium Bitartrate	45 g.
Color	0.2 g.

PUDDING MIXTURES. For every 10-kilogram base use:

FOR VANILLA

Vanillin	4 g.
Powdered Sugar	50 g.

FOR LEMON

Lemon Oil	10 g.
Magnesia Carbonate	5 g.
Powdered Sugar	50 g.

FOR ALMOND

Bitter Almond Oil	5 g.
Magnesia Carbonate	5 g.
Powdered Sugar	50 g.

FOR PINEAPPLE

Pineapple Ether	5 g.
Magnesia Carbonate	5 g.
Powdered Sugar	50 g.

FOR CHOCOLATE

For every 50-gram base: 30 grams of cocoa.

Directions for making pudding: Take 100 grams and ½ liter of boiling milk. Incorporate 7 tablespoonsful of the powder into some of the milk (cold). Bring the rest of the milk to a boil, add cold mixture and continue boiling and stirring until stiff.

VANILLA SAUCE (For the above puddings)

Powdered Sugar	30 g.
Starch	6 g.
Vanillin	0.02 g.
Tartrazin	0.015 g.

Directions: Use with 200 grams of milk and boil up.

CHAPTER VI

BAKERY PRODUCTS, SPECIALTY FOODS, PREPARED CAKE MIXES

The formulae and methods of manufacturing the following products will lay the foundation of this highly specialized business. These formulae may, in some instances, be manufactured and sold to the baker, or the baker may make them for his own use.

The success of a formula rests with the individual trying to make the product. There are many reasons for failures. Lack of experience is one; inferior raw materials is another; improper equipment is still another. Practice alone will make perfect.

POWDERS

MARSHMALLOW POWDER

Gelatin	50 lb.
XXXX Sugar	40 lb.
Albumen	10 lb.

DIRECTIONS FOR USE

I. For filling Cream Rolls, Layer Cakes, and Tarts.

Dissolve 5 ounces of marshmallow powder in 1 quart of warm water and beat up.
Add 3 pounds of granulated sugar and beat until good and stiff.
Flavor with vanilla.

Beat in medium speed after sugar is added. If too stiff a little warm water may be added.

PRECAUTIONS

All utensils must be scrupulously clean and free from grease and dirt.

Have the kettle and beater warm, using hot water. Beat in a warm room, as the marshmallow tends to stiffen in cool atmosphere and you cannot beat properly.

II. For Marshmallow Icings and Toppings

Dissolve 5 ounces of marshmallow powder in 1 quart of warm water and beat up.

Add 3 pounds of granulated sugar and beat until good and stiff. Add 1 pound of corn syrup (glucose) to advantage. Flavor with vanilla.

Beat in high speed.

MARSHMALLOW POWDER

This product stiffens up very quickly.

Gelatin	18 lb. 12 oz.
Cornstarch	9 lb. 6 oz.
Gum Arabic	3 lb. 2 oz.
Albumen	6 lb. 4 oz.
XXXX Sugar	7 lb. 13 oz.
Alum U.S.P.	4 lb. 11 oz.

MARSHMALLOW POWDER

Albumen	40 lb.
Gelatin	40 lb.
XXXX Sugar	40 lb.
Cornstarch	75 lb.
Gum Arabic	12½ lb.
Salt	½ lb.

MERINGUE POWDER

A highly recommended WHIP POWDER for Toppings, Cream Rolls, Kisses, Meringues and all kinds of pastries.

Albumen	10 lb.
Cornstarch	4½ lb.
XXXX Sugar	6 lb.
Gelatin	2 oz.
Alum U.S.P.	1½ lb.
Salt	½ lb.

(If desired the ALUM can be replaced by PECTIN.)

Directions for use: Take 1 quart of cold water and sift 6 ounces of meringue powder on top. Beat well. Then add gradually 3 pounds of granulated sugar and whip until stiff. Flavor with vanilla.

For kisses use 7 pounds instead of 3 pounds of granulated sugar.

MERINGUE POWDER (High-Grade)

Albumen	130 lb.
Tapioca Flour	42 lb.
XXXX Sugar	20 lb.
Alum U.S.P.	4 lb.
Gelatin	2 lb.
Vanillin	3 oz.
Salt	2 lb.

MERINGUE POWDER

Albumen	30 lb.
XXXX Sugar	25 lb.
Gelatin	5 lb.
Gum Arabic	3 lb.
Alum U.S.P.	1 lb.
Indian Gum	½ lb.
Citric Acid	1½ lb.

Note: There are quite a number of meringue or whip powders on the market which contain SAPONIN or SOAP BARK. These preparations, while they whip up to a large volume, are injurious to health and are forbidden by the Pure Food Law.

MERINGUE POWDER FOR LEMON PIE

Powdered Egg Albumen	4 oz.
Granulated Sugar	14 oz.
Salt	¼ oz.
Tapioca Starch	½ oz.
Lemon Powder	¼ oz.

Place 2 pints of water in a thoroughly clean and dry bowl. On heat speed, add gradually the above preparation, until the meringue comes to a good peak.

EGG WHITE COMPOUND

An egg white mixture for boiled icings.

Albumen	60 lb.
Gelatin	10 lb.
Standard Powdered Sugar	50 lb.

DIRECTIONS

For Boiled Icing (Boil to a thin thread or about 245° F.)

Granulated Sugar	16 lb.
Water	1¾ qt.

Dissolve

Egg White Compound	10 oz. in
Water	1 qt.

Place in beater and beat 3 minutes at fast speed. Then add boiled sugar and corn syrup, 1 to 1½ pounds, and vanilla extract.

CAKE WHITE COMPOUND

For Angel Food Cake, White Box Cakes, Layer Cakes, White Loaf Cakes.

Albumen	131 lb.
Tapioca Flour	73½ lb.
Vanillin	3 oz.

Directions: Dissolve 6 ounces of cake white compound in 1 quart of cold water. Beat up and add 1½ pounds of XXXX sugar. Sift together 10 ounces of cake flour, 10 ounces of cornstarch with ½ ounce of cream of tartar. Mix in lightly after eggs and sugar are beaten. Flavor with salt and vanilla.

EGG WHITE

A cheap egg white compound.

Albumen	4½ lb.
Dextrin (White)	2¾ lb.
Sugar (Powdered)	1¾ lb.
Gelatin	1 oz.
Alum U.S.P.	¾ lb.
Salt	¼ lb.

ICING POWDER

An economical product for making various icings.

Gelatin	87½ lb.
XXXX Sugar	87½ lb.
Gum Arabic	22 lb.

Directions: Put 1 ounce of icing powder in a clean pan or vessel, add to this ½ pint of hot water. Stir until dissolved. It is then ready for use. Use as you would egg whites. If you do not use all the liquid at one time, bottle the remainder and it will keep 5 days.

Can be colored and flavored to produce any icing.

ICING POWDER

A cheaper product.

Sugar	9 lb.
Gum Karaya	1 lb.
Tartaric Acid	4 oz.

PIE THICKENER

Powdered Sugar	50 lb.
Gum Karaya	30 lb.
Gelatin	2 lb.
Tapioca	18 lb.

KREMAID (Ice-Cream Powder) or PIE THICK (Pie Thickener)

XXXX Sugar	20 lb.
Gum	20 lb.
Sago Flour	20 lb.

PIE THICK (For Bakers)

Pie Thick is an excellent thing for custard pies. It immediately thickens the custard, not allowing the cornstarch or other ingredients to settle on the bottom crust. It keeps the bottom crust dry and flaky, the custard will not break in transportation, will not be leathery, will stand up, never falling down in the center, and besides Pie Thick will save from four to eight eggs to every gallon of custard.

An excellent custard is made with Pie Thick as follows:

To 1 gallon of milk use 1½ pounds of granulated sugar, 4 ounces of cornstarch, 12 to 16 whole eggs, a good pinch of salt, pinch of mace and 1 ounce of Pie Thick. Sugar, cornstarch, salt, mace, and Pie Thick are mixed together dry, beat eggs and half of milk good, then add dry mixture and rest of milk and beat to a finish.

A cheaper custard than the above can be made by using less eggs and a little more Pie Thick.

To use for cream puff or eclair filling, simply bring above mixture to boil and flavor.

A little Pie Thick added to meringue will help it beat up and it will not become watery.

FOR ICE-CREAM MAKERS

Directions: Take ¾ ounce of Kremaid to each gallon of raw cream or 1½ to 2 ounces of Kremaid to each gallon of raw milk (according to richness of the milk). Mix the proper amount of Kremaid dry with the sugar to be used and pour on it enough of the cream or milk to make a stiff paste. When mixed with a paddle into a paste add rest of liquid, slowly stirring all the time. Flavor and freeze.

Caution: Do not strain after mixing as a lot of the Kremaid will stay in the strainer.

Kremaid requires no cooking or soaking and is always ready for use.

Where Kremaid is used ice-cream will shrink less when transferred from one vessel to another than when made in any other manner or with any other thickener.

PIE THICKENER (Pie Improver for Fruit Pies)

Changes the juices into a jelly-like consistency, increasing thereby the volume and developing the natural fruit flavor.

Indian Gum	10 lb.
XXXX Sugar	20 lb.

Directions for use: To 1 quart of fruit juice from a No. 10 can add the equal amount of water. Mix 4 ounces of pie improver with enough sugar to suit taste (about 1 pound), bring to a boil and add the fruit. Let cool before filling the pies. When using dried fruit use 2 ounces of pie thickener to the quart.

WHOLE EGG POWDER

(1 pound is equal to 40 shell eggs.)

Egg Yolk	80 lb.
Albumen	20 lb.

Directions: To 1 pound of whole egg powder add 3 pints of cold water. Let stand overnight. Stir before dipping out each time, as solids come to the top.

WHOLE EGG POWDER

COMPOUNDS

Powdered Egg Albumen	10 lb.
Powdered Spray Yolk	5 lb.
Gum Karaya	15 lb.
Rice Flour	20 lb.

To improve, add 2 pounds more egg yolk to each 10 pounds.

WHOLE EGG POWDER

No. 1

Egg Yolk No. 1	76 lb.
Egg Yolk No. 2	12 lb.
Albumen	44 lb.

No. 2

Egg Yolk No. 2	88 lb.
Albumen	42 lb.

WHOLE EGG POWDER

COMPOUND

Egg Yolk	70 lb.
Powdered Sugar	14 lb.
Powdered Albumen	8 lb.
Powdered Starch	6 lb.
Bicarbonate of Soda	8 lb.
Cream or Tartar	1 lb.

Or

Egg Yolk	5½ lb.
Albumen	½ lb.
Powdered Milk	4 lb.
Bicarbonate of Soda	1 lb.

Or

Egg Yolk	70 lb.
Albumen	8 lb.
Sugar	14 lb.
Bicarbonate of Soda	8 lb.
Arrowroot	6 lb.

Or

Egg Yolk	70 lb.
Potato Starch	3 lb.
Powdered Sugar	10 lb.
Albumen	10 lb.
Starch	3 lb.
Salt	1 lb.
Bicarbonate of Soda	7 lb.
Cream of Tartar	1 lb.

Or

Egg Yolk	67 lb.
Albumen	18 lb.
Potato Starch	15 lb.

HOT CHOCOLATE (Prepared Cocoa)

Powdered Milk	13 lb.
Sugar	25 lb.
Cocoa	27 lb.
Vanillin	1 oz.
Salt	1 oz.
Ground Cinnamon	1 oz.

Directions: Take 1 heaping teaspoonful of hot chocolate and put in glass. Add small portion of boiling water (not hot water). Stir well with spoon into a smooth paste. Then add the balance of boiling water to make a glass full.

LEMONADE POWDER

Citric Acid	1½ lb.
Granulated Sugar	5 lb.
Terpeneless Lemon Oil* or Terpeneless Orange Oil	¼ fl. oz.

POUND CAKE FLAVOR

Corn Meal Flour	10 lb.
Granulated Egg Yolk	3 lb.
Rice Flour	6 lb.
Peanut Oil	1 pt.
Mace Oil (Or Nutmeg Oil)	1 oz.

*For Cherry, Strawberry, Raspberry, etc., use the synthetic Fruit Oils F, say 6–8 fluid ounces.

Or

Tapioca Flour	5 lb.
Sugar	5 lb.
Mace (Ground)	¼ oz.
Nutmeg (Ground)	¼ oz.
Apple Pie Flavor	2 oz.
Lemon Oil	¼ oz.
Orange Oil	¼ oz.
Vanillin	¼ oz.

APPLE PIE FLAVOR

Tapioca Flour	25 lb.
XXXX Sugar	25 lb.
Cassia Oil	13 oz.

CAKE MOIST

Tapioca Flour	10 lb.
Starch	10 lb.
Powdered Milk	10 lb.

MALT MILK

Malt Extract	10–12 lb.
Dry Milk Powder	88–90 lb.

PIE CRUST

Pastry Flour (Winter Wheat Flour)	3 cups
Sift with	
Baking Powder	1 full teaspoon
Salt	1 level teaspoon

Directions: Mix the "Pie Crust" with 1 cup of lard or other good shortening and make to a dough with ½ cup of freshly boiling water. Be sure to cut holes with a fork in the upper crust for steam to escape.

Bake the pie in a hot oven for about ½ hour.

POWDERED FUDGE

Sugar	1½ oz.
Dry Skimmed Milk	⅛ oz.
Cocoa	20 gr.
Gum Tragacanth	10 gr.
Vanillin and Salt	Trace

Directions: Melt 1 ounce of butter with 4 cubic centimeters of water. Add the above quantity of powdered fudge and stir until creamy. Let cool.

LEMON POWDER FOR FLAVORING

LEMON SUGAR

Lemon Oil	4 oz.
Magnesia Carbonate	4 oz.
Powdered Sugar	4 lb.
Dextrin (Yellow)	1 lb.

Or

Granulated Sugar	12 lb.
Tartaric Acid	12 oz.
Lemon Essence	8½ oz.

VANILLA POWDER FOR FLAVORING

VANILLA SUGAR

Vanillin	1 oz.
Powdered Sugar	5 lb.
Powdered Licorice	½ oz.
Mix	

BUTTER POWDER

Sodium Bicarbonate	4 lb.
Sodium Chloride	1 lb.
Orange I	10 gr.

EGGY (Egg Substitute)

Cornstarch	8 oz.
Bicarbonate of Soda	8 oz.
Cream of Tartar	5 oz.
Powdered Turmeric	½ oz.
Casein	3 oz.
Albumen	2 oz.

Directions: Beat a level teaspoonful with a tablespoonful of hot water and use as an egg.

PIE FILLING POWDERS

LEMON PIE FILLING POWDER

No. 1

Lemon Juice Powder	16 lb.
Whole Egg Powder	10 lb.
Cornstarch	32 lb.
Tapioca Starch	8 lb.
Gelatin	½ lb.
Citric Acid	4 oz.
Lemon Oil	1 oz.
Indian Gum	1 lb.
Egg Color	¼ oz.

No. 2

As No. 1 but use	
Lemon Juice Powder	10 lb.

No. 3

Cornstarch	125 lb.
Ground Tapioca Flour	25 lb.
Citric Acid	20 oz.
Gelatin	18¾ oz.
Albumen	1 lb.
Lemon Oil	16 oz.
Terpeneless Lemon Oil	1 oz.
Orange Peel	20 oz.
Indian Gum	6 oz.
Tartrazin	1 oz.

Type of Label

> COMPOUND PIE FILLING (Dry) No. 1
>
> Lemon Flavor—Artificial Color
>
> DIRECTIONS: Put 2½ quarts of water on the fire and bring to boil. Dissolve 1 pound of powder and 2½ pounds of sugar in 1 quart of water, and add to boiling water on fire. Pour gradually, stirring continually. Bring to a boil again, and add ½ to 2 ounces of butter or lard. Allow to cool before using. Meringue with Meringue Powder.

LEMON PIE FILLING POWDER (C)

Cornstarch	75	lb.
Tapioca	20	lb.
XXXX Sugar	5	lb.
Lemon Oil	7	oz.
Egg Color	1	oz.
Salt	1	lb.
Granulated Egg Yolk	10	lb.
Lemons to 100 lb.	10	
or		
Lemon Juice Powder (Dry)	3½	lb.
or		
Lemon Juice Powder (Dry)	1½	lb.
Citric Acid	1	lb.
Lemon Oil	1	lb.
Tartrazin (Instead of Egg Color)	3	dr.

LEMON PIE FILLING POWDER (H)

Cornstarch	21	lb.
XXXX Sugar	4½	lb.
Whole Egg Powder	½	lb.
Indian Gum	1	oz.
Salt	1½	oz.
Terpeneless Lemon Oil	½	oz.
Citric Acid	½	oz.
Lemon Color	⅜	oz.

Type of Label

> COMPOUND PIE FILLING (Dry)
>
> Lemon Flavor—Artificial Color
>
> DIRECTIONS: Place 4 quarts of water and 3 pounds of sugar on fire and bring to a boil.
>
> Dissolve 1 pound of powder in 1 quart of cold water and add to the above solution.
>
> Remove from fire in 2 to 3 minutes.

LEMON PIE FILLING POWDER (Cheap)

Cornstarch	60	lb.
Citric Acid	1	lb.
Lemon Oil	1	lb.
Lemon Color	⅜	oz.

INSTRUCTIONS FOR USE OF THE ABOVE LEMON POWDER

Boil 3 quarts of water with 3 pounds of sugar. Mix 1 quart of Lemon Powder in 1 quart of cold water; pour cold mixture into boiled water and sugar mixture and stir briskly. Remove from fire as soon as mixture is properly thickened and let cool.

CREAM PUFF FILLER (Vanilla Cream or Cream Pie Filler)

No. 1

Cornstarch	60	lb.
Tapioca Flour	10	lb.
Gelatin	½	lb.
Albumen	½	lb.
Egg Yolk, Granulated	3	lb.
Whole Dry Milk	5	lb.
Salt	½	lb.
Vanillin	1	oz.
Special Vanilla Flavor *	4	oz.
Egg Color	¼	oz.
Indian Gum	1	lb.

No. 2

As No. 1 but use

Egg Yolk	2	lb.
Whole Dry Milk	2	lb.

* Special Vanilla Flavor	Vanillin	4 oz.
	Coumarin	1 oz.
	Alcohol	1 pt.
	Warm Water	1 pt.

CREAM PUFF FILLER (C)

Milk Powder (Skim)	20	lb.
XXXX Sugar	20	lb.
Cornstarch	110	lb.
Salt	2	lb.
Vanillin *	3	oz.
Coumarin *	½	oz.
Glycerin	1	pt.
Egg Color	1	oz.
Egg Yolk, Granulated	4	lb.

Directions: Place 3 quarts of water and 3 pounds of sugar on the fire and bring to a boil.

Dissolve 1 pound of powder in 1 quart of water and add to the above solution. Stir. Remove from fire as soon as it thickens up.

* Dissolve by heat in the glycerin.

CREAM PUFF FILLER (H) (Sweet)

Cornstarch	10	lb.
Dried Skimmed Milk	2½	lb.
Egg Yolk, Granulated	¾	lb.
Albumen	¼	lb.
Sugar	14	lb.
Gelatin or Gum	1	oz.
Vanillin	¼	oz.
Coumarin	1/16	oz.
Egg Color	⅜	oz.

CHOCOLATE PIE FILLER

Cornstarch	48	lb.
Flour (Tapioca)	16	lb.
Egg Yolk, Granulated	6	lb.
Milk Powder	6	lb.
Salt	12	oz.
Cocoa	12	lb.
Vanillin	1	oz.

CHOCOLATE PIE FILLER (C)

Cocoa	40	lb.
Cornstarch	97.5	lb.
XXXX Sugar	5	lb.
Salt	1	lb.
Vanillin	1	oz.
Indian Gum	1	lb.
Tapioca Flour	17	lb.
Chocolate Color	½	oz.

CHOCOLATE PIE FILLER (H)

Cocoa	30	lb.
Cornstarch	65	lb.
XXXX Sugar, Powdered	5	lb.
Salt	3	oz.
Vanillin	1	oz.
Indian Gum	1	lb.

Directions: One pound of filler, 2½ pounds of sugar, 4 quarts of water. See directions under Lemon Pie Filling Powder.

CARAMEL CUSTARD PIE FILLER

To 10 pounds of Pie Base add:

Roasted Powdered Whole Milk	1	lb.
Brown Sugar	2½	lb.
Caramel Flavor	5	fl. oz.
Chocolate Color	¼	oz.

Directions: Take 1 pound of caramel custard pie filler and add 2 pints of water, mixing thoroughly with an egg whip. Now put on fire 2 pints of water and add 12 ounces of brown sugar and bring to a boil. Pour the other mixture into this, stirring continually until it thickens.

Pour into shells and bake same as custard pies.

To increase the caramel taste, put 3 or 4 tablespoonfuls of granulated sugar on the fire with a little water and allow it to cook to a golden brown. Then pour a little into each pan before filling custard.

BUTTERSCOTCH PIE FILLER

To 10 pounds of Pie Base add:

Roasted Powdered Whole Milk	1½ lb.
Brown Sugar	2½ lb.
Butterscotch Flavor	1 oz.

Prepare as one of the other pie filling powders.

CHERRY CREAM PIE FILLER

Whole Egg Powder	2 lb.
Granulated Sugar	8 lb.
Dry Skim Milk	6 lb.
Cornstarch	2 lb.
Salt	1 oz.
Pie Thickener	5 oz.
Cherry Flavor, Concentrated True Fruit	1 oz.
Color	¼ oz.

BASE FOR COMPOUND DRY PIE FILLING POWDERS

CHERRY, STRAWBERRY, RASPBERRY, ETC.

Cornstarch	47½ lb.
Tapioca Flour	21 lb.
Tartaric Acid Powder	8¾ oz.
Gelatin	8¾ oz.
Indian Gum	3½ oz.
Salt	10½ oz.
Albumen	8¾ oz.
True Fruit Flavor Concentrate	8¾ oz.
Color	To suit

Directions: Place 3 quarts of water and 3 pounds of sugar on the fire and bring to a boil. Dissolve 1 pound of powder in 1 quart of warm water and add to the above solution. Remove from fire as soon as it thickens. Meringue with meringue powder.

POWDERED MALTED MILK PIE FILLER

Skim Milk Powder	5 oz.
Granulated Sugar	18 oz.
Powdered Egg Yolk	2 oz.
Cornstarch	3 oz.
Salt	⅛ oz.
Malted Milk Powder	5 oz.
Dark Cocoa Powder	3 oz.
Flavor	To suit

Add the above to 2 pints of boiling water and stir to a thick consistency.

PREPARED POWDERED CAKE MIXES

SPICY CAKE MIX (HONEY CREAM CAKE MIX)

Winter Wheat Flour	8 lb.
Ground Cinnamon	1½ lb.
Cream of Tartar	2 lb.
Bicarbonate of Soda	3 lb.
Ground Mace	½ lb.
Ground Nutmeg	½ lb.
Ground Cardamom	½ lb.
Ground Ginger	½ lb.
Ground Cloves	½ lb.

RECIPE

Rub up

Lard or Compound	¾ lb.
Granulated Sugar	3¾ lb.

Add

Eggs	5 lb.
Spicy Cake Mix	1 lb.
Salt	1 oz.

Gradually add

Sour Milk or Buttermilk	3 qt.
Cake Flour	7 lb.
Ground Raisins	2 lb.

Bake in slow oven. This will make layer, sheet, and loaf cake; also cup cake and muffins. A 2-layer, 7-inch cake with icing will cost about 10 cents to make. A trade winner.

DEVIL'S FOOD MIX

Brown Sugar	15 lb.
Bicarbonate of Soda	6 lb.
Cream of Tartar	4 lb.
Winter Wheat Flour	30 lb.
Dutch Cocoa	15 lb.
Ground Cinnamon	4 oz.
Vanillin	3 oz.
Powdered Buttermilk	7½ lb.
Chocolate Color	4½ oz.
Amaranth	1½ oz.

If a very dark cocoa can be purchased, it is desirable to cut the amount of color.

RECIPE

Cream up

Sugar, Granulated	2½ lb.
Shortening	1 lb.
Eggs	1 pt.

Now add

Devil's Food Mix	2 lb.
Salt	1 oz.

Mix thoroughly, then add

Flour	2¼ lb.

Thin down with water (2 to 2½ pints) to desired consistency. Bake in temperature of 350° to 375° F.

SPONGE CAKE MIX

XXXX Sugar	6 lb.
Granulated Egg Yolk	1½ lb.
Skim Milk Powder	1 lb. 14 oz.
Baking Powder	7 oz.
Albumen	8 oz.
Winter Wheat Flour	6 lb.
Salt	1½ oz.
Lemon Oil	15 drops
Egg Color	Trace

CHOCOLATE SPONGE CAKE MIX

As Sponge Cake Mix but use

Cocoa	¾ lb.
and	
Flour	5¼ lb.

Nothing to add but water. Will produce any form of Sponge Cake, Sponge Layers, Sheet Layers, Loaf Cakes, Lunch Cakes, Jelly Rolls, etc.

Directions: Take 12 ounces of lukewarm water and mix with 2 pounds of Sponge Cake Mix and make dough. Then put in machine and beat for 4 minutes. Scale into pans, bake in a moderate oven.

SUGAR COOKIE MIX

XXXX Sugar	3 lb.
Winter Wheat Flour	6 lb.
Cottonseed Oil	6 oz.
Bicarbonate of Soda	1 oz.
Egg Yolk, Granulated	1 oz.
Whole Milk Powder	5 oz.
Salt	1½ oz.
Egg Color Dandy Cake Flavor	Trace

SPICY COOKIE MIX

Brown Sugar	3 lb.
Winter Wheat Flour	6 lb.
Cottonseed Oil	6 oz.
Bicarbonate of Soda	1 oz.
Egg Yolk, Granulated	1 oz.
Buttermilk Powder	5 oz.
Salt	1½ oz.
Cinnamon	1 oz.
Allspice	¼ oz.
Ginger	¼ oz.
Nutmeg	¼ oz.

A preparation that will produce quickly and at a low cost the most delicious cookies.

Directions: Mix 5 pounds of Cookie Mix with 1 quart of water and make dough. Roll out and cut.

Two pounds of Cookie Mix makes 3½ to 4 dozen cookies.

YELLOW LAYER CAKE

Cake Flour	5 lb.
Baking Powder	5 oz.
Powdered Whole Milk	1 lb.
Granulated Sugar	6½ lb.
Shortening	2 lb.
Powdered Egg Yolk	1 lb.
Salt	2 oz.
Vanilla Extract	1 oz.
Cornstarch	8 oz.

Sift together flour, baking powder, and milk. Cream the sugar with the shortening and add the egg. Carefully mix all of the ingredients. This mixture will require about 5 pounds of water.

Bake in oven at 420° F. for about 15 to 20 minutes.

DEVIL FOOD LAYERS

Cake Flour	5 lb.
Baking Powder	3 oz.
Powdered Whole Milk	1 lb.
Granulated Sugar	5 lb.
Shortening	2 lb.
Egg Albumen	4 oz.
Gum Tragacanth	2 oz.
Skim Milk	1 lb.
Cocoa Powder	1 lb.
Baking Soda	2 oz.
Salt	2 oz.

Sift together all but the sugar and shortening, which should be creamed and then added to the sifted mixture. This mixture will require about 7 pounds of water.

Bake in oven heated to about 400° F.

WHITE POUND CAKE

Cake Flour	5 lb.
Powdered Whole Milk	1 lb.
Granulated Sugar	5 lb.
Shortening	2½ lb.
Powdered Egg Albumen	4 oz.
Gum Tragacanth	2 oz.
Vanilla Extract	1 oz.
Salt	1 oz.
Baking Powder	3 oz.

Cream shortening and sugar, then add to the other ingredients, which have been sifted all together. Mix carefully. This mixture will require about 5 pounds of water.

Bake in a slow oven at 300° F. to 325° F.

YELLOW POUND CAKE

Cake Flour	5	lb.
Powdered Whole Milk	10	oz.
Granulated Sugar	6	lb.
Shortening	3	lb.
Powdered Egg Yolk	1½	lb.
Powdered Mace	3	g.
Salt	2½	oz.

Cream shortening and sugar, then add to the other ingredients, which have been sifted all together. Mix thoroughly. This mixture will require about 4 pounds of water.

Bake 1 hour in a moderate oven at about 350° F.

LEMON CUP CAKES

Cake Flour	5	lb.
Powdered Whole Milk	1	lb.
Granulated Sugar	6	lb.
Baking Powder	5	oz.
Shortening	2	lb.
Powdered Egg Yolk	12	oz.
Powdered Egg Albumen	½	oz.
Powdered Lemon Juice	7	oz.
Salt	1½	oz.

Cream sugar and shortening. Sift all other ingredients and mix well together. Then work in about 5 pounds of water and mix until smooth and creamy. Fill into baking cups and bake in an oven at 375° F. to 400° F. for about 20 minutes. Use a Lemon Fondant Icing.

HIGH RATIO CHOCOLATE MALTED MILK CAKE

Cake Flour	5	lb.
Skim Milk Powder	1½	lb.
Malted Milk Powder	1½	lb.
Granulated Sugar	7	lb.
Shortening	2½	lb.
Baking Powder	3½	oz.
Baking Soda	¾	oz.
Powdered Egg Yolk	12	oz.
Vanilla Extract	½	oz.
Dark Cocoa Powder	2	lb.
Salt	3	oz.

Cream sugar and shortening, then add to the rest of the ingredients, which have been sifted together, and mix thoroughly. This mixture will require about 10 to 12 pounds of water to make a smooth creamy dough.

Bake at 375° F.

HIGH RATIO SPICE CAKE

Cake Flour	5	lb.
Skimmed Milk Powder	10	oz.
Granulated Sugar	3½	lb.
Brown Sugar	3	lb.
Powdered Molasses	1	lb.
Shortening	3	lb.
Powdered Egg Yolk	1	lb.
Powdered Egg Albumen	1	oz.
Baking Powder	5	oz.
Salt	2	oz.
Cinnamon	50	g.
Cardamom	20	g.
Cloves	10	g.
Nutmeg	10	g.
Ginger	10	g.

Cream shortening and sugar, then add to the other ingredients, which have been sifted all together. This mixture will require about 7 pounds of water.

Bake in oven at about 375° F.

HIGH RATIO DEVIL'S FUDGE CAKE

Cake Flour	4	lb.
Skimmed Milk Powder	12	oz.
Sugar	7	lb.
Dark Cocoa Powder	1	lb.
Powdered Egg Yolk	1¼	lb.
Powdered Egg Albumen	1	oz.
Baking Powder	3	oz.
Baking Soda	1	oz.
Shortening	2½	lb.
Salt	3	oz.
Vanilla Extract	To	taste
Cinnamon	½	oz.

Carefully mix and sift the above ingredients. The amount of water needed for the proper dough consistency is about 8 pounds. Mix until smooth. This formula may be used for layer cake, cup cakes, loaf cakes, etc.

Bake in oven at about 375° F.

HIGH RATIO PLATINUM WHITE CAKE

Cake Flour	5	lb.
Skimmed Milk Powder	12	oz.
Granulated Sugar	7	lb.
Shortening	2	lb.
Powdered Egg Albumen	4	oz.
Gum Tragacanth	10	oz.
Baking Powder	5	oz.
Cream of Tartar	1	oz.
Vanilla Extract	3	oz.
Salt	3	oz.

Carefully mix and sift all the ingredients until a uniform powder. Add about 7 pounds of water and mix until smooth and creamy. This formula may be used for layer cakes, cup cakes, ring cake, sheet cake, and loaf cakes.

Baking temperature depends upon the form of cake to be baked.

HIGH RATIO GOLDEN CAKE

Cake Flour	5 lb.
Skimmed Milk Powder	14 oz.
Sugar	7 lb.
Shortening	2 lb.
Powdered Egg Yolk	12 oz.
Baking Powder	4 oz.
Salt	3 oz.
Vanilla Extract	3 oz.

Carefully mix these ingredients into a uniform powder. Add about 8 pounds of water and mix thoroughly until a smooth consistency is formed.

Bake in oven at 375° F.

This formula may be used for strawberry shortcake, jelly rolls, cup cakes, layer cakes, etc.

POUND CAKE FLAVOR

Cornmeal	10 lb.
Granulated Egg Yolk	3 lb.
Rice Flour	6 lb.
Peanut Oil	1 pt.
Mace Oil	1 oz.

POUND CAKE FLAVOR

Mace	¼ oz.
Nutmeg	¼ oz.
Apple Pie Flavor	2 oz.
Sugar	5 lb.
Tapioca Flour	5 lb.
Lemon Oil	¼ oz.
Orange Oil	¼ oz.
Vanillin	¼ oz.

MIXED SPICES

A. CAKE SPICES

For fruit and spice cakes.

Cinnamon, Ground	1 lb.
Ginger, Ground	¼ oz.
Cloves, Ground	4 oz.
Mace, Ground	1 oz.

APPLE PIE SPICE

Mace, Ground	½ lb.
Cinnamon, Ground	1½ lb.
Cayenne Pepper	1¼ lb.

If mace is high, use nutmeg.

MINCE MEAT SPICE

Cinnamon	67½ lb.
Allspice	22½ lb.
Cloves	22½ lb.
Mace	14½ lb.

PUMPKIN PIE FLAVOR (Furnishes the Brown Crust)

Ginger	1 lb.
Cinnamon	1 lb.
Salt	1¼ lb.
Mace (Nutmeg)	½ lb.
Cloves	¼ lb.

Directions: Add no salt to the mix. For each gallon of mix use 4 ounces of molasses and 2½ ounces of the pumpkin pie flavor.

POUND CAKE FLAVOR SPICE

Nutmeg	30 lb.
Colored Cereal (Cornmeal)	60 lb.

PREPARED FLOURS

WAFFLE FLOUR

This prepared WAFFLE FLOUR will produce waffles of fine texture, taste, and quality.

Directions: Mix 1 pound of waffle flour with 10 fluid ounces of water. Grease the iron and bake until golden brown.

Sugar, Powdered	3 lb.
Cottonseed Oil	1 lb.
Whole Egg Powder	12 oz.
Winter Wheat Flour	8 lb.
Baking Powder	6 oz.
Salt	2 oz.

Incorporate the oil into the flour, follow with the egg powder, sugar, and the salt. Add the baking powder last. The oil and flour are best prepared in a mixing machine and, when dry enough, transferred to a powder mixing machine, provided with a sifter on top (Day Mixer).

DOUGHNUT FLOUR

A scientifically prepared DOUGHNUT FLOUR that produces the most palatable and appetizing doughnuts. Can be used for machine and hand work. Nothing to add but water, when made by hand. For machine work add 1 to 2 eggs to 5 pounds of doughnut flour.

Directions: Mix 5 pounds of doughnut flour with 1 quart of ice cold water. Roll out and cut.

Makes 10 dozen doughnuts.

Fry at a temperature of 385° to 400° F.

DOUGHNUT FLOUR

Formula No. 1

Winter Wheat Flour	98 lb.
Baking Powder	3 lb. 2 oz.
Cream of Tartar	12 oz.
Bicarbonate of Soda	6 oz.
Cottonseed Oil	4 lb.
XXXX Sugar	25 lb.
Egg Yolk, Granulated	2 lb.
Skimmed Powdered Milk	5 lb.
Lemon Oil	½ oz.
Mace Oil	½ oz.
Salt	20 oz.

DOUGHNUT FLOUR

No. 2

Sugar	20 lb.
Shortening	3½ lb.
Mace Oil	⅛ oz.
Cinnamon Oil	10 drops
Salt	1 lb.
Vanillin	To suit
Powdered Egg	2 lb.
Skim Milk	5 lb.
Cake Flour	61 lb.
Baking Powder	3 lb.

Use fine granulated sugar. Cream the shortening with the sugar, and add the spice oils. Beat up egg with sugar, and add some creamed shortening. Then add this to the creamed sugar. Sift the cake flour, milk powder, and baking powder and add to creamed sugar. Put through a mixer and sifting machine.

DOUGHNUT FLOUR

No. 3

Sugar	29 lb.
Shortening	3½ lb.
Mace Oil	⅛ oz.
Salt	1 lb.
Vanillin	To suit
Powdered Egg	4 lb.
Skim Milk	4 lb.
Cake Flour	57 lb.
Baking Powder	3½ lb.

Mix the same as suggested above for formula No. 2.

To make doughnuts use 16–17 pounds of doughnut flour with 3¼ quarts of water. Baking temperature around 390° F.

POWDERED DOUGHNUT SUGAR

Pulverized Dextrose	75 lb.
Cake Flour	10 lb.
XXXX Sugar	9 lb.
Shortening (Around 96° F. Melting Point)	5 lb.
Vanillin	1 oz.

Melt the shortening and cream with some of the dextrose sugar. Gradually work in the balance of the dextrose, and when thoroughly incorporated add the balance of the ingredients. Put the entire mix through a mixer and sifting machine.

MILK BISCUIT FLOUR

Winter Wheat Flour	11 lb.
Whole Milk Powder	10 oz.
Mono Calcium Phosphate	7 oz.
Bicarbonate of Soda	3¼ oz.
Salt	4 oz.

SELF-RAISING FLOURS

BISCUIT FLOUR

Mono Calcium Phosphate	7 lb.
Bicarbonate of Soda	3¼ lb.
Fine Salt	3 lb.
Winter Wheat Flour	186¾ lb.

PANCAKE FLOUR

Mono Calcium Phosphate	7 lb.
Bicarbonate of Soda	3¼ lb.
Fine Salt	3 lb.
Kiln-dry White Corn Flour	100 lb.
Medium Grade Wheat Flour	86¼ lb.

BUCKWHEAT FLOUR

Mono Calcium Phosphate	7 lb.
Bicarbonate of Soda	3¼ lb.
Fine Salt	3 lb.
Buckwheat Flour	130 lb.
Low Grade Wheat Flour	56¾ lb.

POUND CAKE FLOUR

A prepared pound cake compound for quick and excellent use. Nothing but water is required.

Directions: Mix 1 pound of cake flour with ½ pint of cold water. Make dough, put in molds, and bake at 400° F.

Winter Wheat Flour	3 lb.
Egg Yolk, Granulated	½ lb.
Baking Powder	¼ lb.
Milk Powder	10 oz.
Sugar	2 lb. 2 oz.
Salt	1½ oz.
Cottonseed Oil	5 lb.
Vanol (Concentrated Vanilla Flavor)	¼ oz.
Lemon Oil	¼ oz.

PANCAKE FLOUR

Corn Flour	20½ lb.
Buttermilk Powder	5 lb.
Winter Wheat Flour	21½ lb.
Mono Calcium Phosphate	1¾ lb.
Bicarbonate of Soda	13 oz.
Salt	12 oz.

Or

Wheat Flour	7 lb.
Corn Flour	2 lb.
Rice Flour	1 lb.
Sugar Powder	¼ lb.
Milk Powder	½ lb.
Salt	½ oz.
Calcium Acid Phosphate	1½ oz.
Bicarbonate of Soda	1 oz.

Directions for use: make batter with water or milk and cook in regular way.

BUCKWHEAT FLOUR

Buckwheat Flour	55 lb.
Winter Wheat Flour	28½ lb.
Milk Powder	10 lb.
Mono Calcium Phosphate	3½ lb.
Bicarbonate of Soda	1 lb. 10 oz.
Salt	1½ oz.
Buttermilk Powder	3 lb.

COCOANUT FLOUR

To 10 pounds of Cream Puff Filler No. 1 add 6–8 ounces of granulated cocoanut.

APRICOT GLOSS

Apricot Paste	40 lb.
Apricot Pulp	80 lb.
Cane Sugar	250 lb.
Agar-agar*	4 lb.
Water	15 gal.
Citric Acid	4 oz.

Bring the water to a boil, then add 200 pounds of cane sugar, and the fruit with the dissolved citric acid. Cook to 223° F. Transfer to cooling table and work in the solution of agar-agar mixture until uniformly mixed.

FRUIT FILLING

Ground Dates	10 lb.
Fig Jam	10 lb.
Sweetened Apple Sauce	4 lb.
Chopped Nut Meats	3 lb.

Mix the above ingredients.

APRITOPPER

Soak overnight in water to cover agar-agar.

Agar-agar *	2 lb.

The next morning drain off the water, and discard it.

Bring to a boil in a 25-gallon copper kettle:

Sugar	100 lb.
Water	1 qt.
Apricot Pulp **	62½ lb. (10 cans of 6 lb. 4 oz.)

Add the above agar-agar, and bring anew to a boil under good stirring. When thoroughly mixed, shut off steam immediately. Strain under a roller strainer into 30-pound pails.

* Let the agar-agar soak overnight in 10 gallons of water. In the morning transfer to kettle, bring to a boil and add the 50 pounds of sugar. Cook to 223° F., and then mix the batch with the apricot cook on the cooling table.

If the above gloss is too tough add less agar-agar, if too soft add more agar-agar.

* Agar-agar: Malayan name for a seaweed, known also as Ceylon Moss and Jaffna Moss. Agar-agar is employed by bacteriologists for the cultivation of bacteria. In the food industry it serves to gelatinize food products.

Agar-agar in powder is added sometimes to pudding powders (1%) to make the pudding smooth.

** "DEMAND" BRAND APRICOT PULP is mostly used. The imported Spanish pulp, because of its high acid content, interferes with the agar-agar.

Type of Label

APRITOPPER

an ideal topping and cake filling

Guaranteed Pure

DIRECTIONS. Place quantity required in pan and dissolve by heating and use while in liquid form.

Use Apritopper with a brush as a coating or spread with a knife on French pastry, tarts, fruit cake, etc. To save time, on small pieces, apply Apritopper by dipping.

Net Weight 30 lbs.

PIE CRUST

Cake Flour	25 lb.
Shortening	18 lb.
Salt	½ lb.
Water	6 lb.
Skimmed Milk Powder	¼ lb.

Mix the skimmed milk powder with the flour. Then mix all the shortening with the flour. Add the cold water slowly to the batch and mix very little.

GLAZES AND WASHES FOR BAKERY GOODS

Glucose	2 lb.
Water	1 lb.

Bring to a good boil, and apply to pieces as they come from the oven.

PECTIN GLAZE FOR FRUIT TARTS

No. 100 Pectin Powder	2 oz.
Sugar	4½ lb.
Water or Fruit Juice	1¼ qt.
50% Phosphoric Acid	1¼ oz.

Mix pectin with 1 pound of sugar. Bring liquids to boil and add remaining sugar. Shut off steam and stir the sugar until dissolved. Then add the pectin-sugar mixture, stirring it around. Allow liquid to cool, then add the acid. This will give a high gloss to fruit tarts.

CAKE SHINE

Water	4 pt.
Sugar	5 lb.
Gelatin	4 oz.

Bring water to a boil, and dissolve the gelatin in it, and immediately add the sugar. When dissolved shut off steam. Apply this shine to cakes while hot.

PECTIN GLAZE OR WASH FOR DANISH PASTRY, SWEET GOODS, ETC.

No. 100 Powdered Pectin	1 lb.
Powdered Cane Sugar	40 lb.
Corn Syrup	50 lb.
Water	4 gal.

Mix 1 pound of pectin with 5 pounds of sugar in a dry pan. Bring water to a boil and add slowly the pectin-sugar mixture, with stirring. Then add balance of sugar, corn syrup, and cook to 220° F.

This syrup is known as Stock Syrup, and is thinned down by adding 1 pint of water to each gallon, for glazing on pastry, etc. Apply this glaze on pastry as soon as taken from oven. By adding 1 ounce of 50% solution of citric acid to 10 pounds of stock syrup, it makes a glaze for any kind of fruit, upon fruit tarts, fruit cakes, etc. After adding acid solution, pour quickly on cakes and spread with a knife before the formed jelly becomes too thick. Add acid to small portions of stock solution at one time.

GLAZED NUT TOPPING

Pectin Stock Syrup	4 lb.
Butter and Shortening	2 lb.
Salt	½ oz.
Vanillin	To suit
Ground Sliced Nuts	1 lb.

Method: The above mixture is creamed lightly and spread on coffee cake dough, and should be applied to dough before proofing. Makes a nice luster topping.

LACQUER (GLACE) FOR CHOCOLATES AND MARZIPAN

Alcohol	74 Kg.
Gum Benzoin	12.5 Kg.
Gum Sandarac, Purest	25 Kg.
Venice Turpentine	1 Kg.

FRUIT CAKE GLACE

Gum Arabic, Lumps	4 lb.
Are soaked in	
Hot Water	1 gal.

and if completely dissolved, strained.

Add

Syrup	1 pt.
and	
Benzoate of Soda	2 dr.

dissolved in 2 fluid ounces of hot water. If necessary strain again. Flavor with 1 ounce of lemon essence.

ROYAL ICING

Gelatin	¼ lb.

Soak in ½ gallon of lukewarm water for 15 minutes. Place on fire, let come to a boil. Skim off dirt. Add 1 ounce of alum. Cool. Place in bottle. Take even part water, beat, and add pulverized sugar.

NUT TOPPINGS

TOASTED BUTTER TOPPING

Heat slowly to a golden brown tint over a slow fire:

A. Shortening	2 lb.
Butter	2 lb.

Mix and sift together

B. Cake Flour	6 lb.
Cinnamon	3 lb.
Brown Sugar	10 lb.

Now gradually work in the hot butter A, with B, with stirring, and until the mixture turns to soft crumbs.

PECAN COCONUT NUT TOPPING

Mix well

Fine Granulated Sugar	6 lb.
Macaroon Coconut	3 lb.
Chopped Pecans	2 lb.
Salt	½ oz.

Add to the above and mix to a soft paste consistency.
Bring to a boil first.

Water	1 qt.
Shortening	1 lb.

Now spread the above paste evenly and not too thickly on toasting pans, or toast in baking oven, until layer develops a deep golden tint. The layer should be turned once in a while to get uniform heating.

ALMOND BRITTLE TOPPING

Cook to 275° F.

Granulated Sugar	6 lb.
Water	3 pt.
Corn Syrup	2 lb.

Add to above 1 pound of chopped almonds and cook to 310° F. Pour onto pans, not too thick, and when cooled off break up into fine almond brittle, and sift.

SESAME SEED AND PEANUT TOPPING

Cook to 240° F.

Granulated Sugar	3 lb.
Corn Syrup	3 lb.
Water	2 pt.

Add to the above ½ pound of sesame seeds and ¾ pound of chopped peanuts and cook to 300° F. Pour on pans, not too thick, let cool, break up fine, and sift according to size pieces required.

CRUMB TOPPING

Almond Paste	45 lb.
Granulated Sugar	40 lb.
Skimmed Milk Powder	7 lb.
Gum Acacia	5 lb.
Coconut Butter (Melting Point around 98° F.)	3 lb.

The above mixture is mixed very thoroughly, then placed in a revolving heated compartment until the batch becomes toasted. Break up the product and sift to proper size.

BOILED MERINGUE

(1) Water	1 qt.
Sugar	3 lb.
Meringue Powder	6–8 oz.

Whip to a stiff meringue.

(2) Water	1½ pt.
Sugar	2 lb.
Agar	1 oz.

Mix sugar, agar, and boil for 3 minutes.

When the meringue (1) is thoroughly whipped, reduce the machine to second speed, and then add the hot agar mixture (2) slowly so that it is entirely absorbed as poured in.

GLAZE FOR FRESH BERRY AND OPEN FACE PIES

(1) Water	22 pt.
Sugar	18 lb.
Agar	2 oz.
Fruit (Apricot or Peach, sieved)	10 lb.
Salt	3 oz.

Mix the sugar with the agar, and then boil the entire batch for 3 minutes. Then add (2).

(2) Water	1 qt.
Cornstarch	30 oz.
Color and Flavor	If desired

Mix together.

The above when cooked makes up a clear filling and will seal the fruit when spread over.

BASIC CREAM

(1) Water	7 pt.
Sugar	12 lb.
Meringue Powder	20 oz.
Honey	½ pt.
Lemon Emulsion	¼ oz.
Vanilla	3 oz.

Beat the above to a stiff meringue.

(2) Water	3	gal.
Sugar	7	lb.
Agar-Agar	2	oz.
Salt	4	oz.
Cornstarch	3	lb.

Mix the sugar with the agar. Draw off 1 gallon of water and make a paste with the starch. Boil the 2 gallons of water, add the sugar-agar mixture. When clear add the starch paste solution, and cook for 3 minutes. To the stiff meringue above (1) add this hot mixture and beat up until cool enough to handle. This mix may be used to make cream pies, with the addition of color and flavor. It may be mixed also with acid-tasting fruits.

FRUIT BARS

Cake Crumbs	6	lb.
Molasses	1½	qt.
Water	2	qt.

Soak until soft, and then cream up.

Sugar	4	lb.
Malt Syrup	2	lb.
Shortening	3	lb.
Soda	6	oz.
Salt	1	oz.
Ginger	½	oz.
Cinnamon	1	oz.

Mix the two parts and add

Raisins or Currants	5	lb.
Coconut	2	lb.
Flour	11	lb.

Scale off in 12-ounce pieces and roll out in long strips. Lay four strips on pan, flatten, and bake in medium oven. These bars may be iced with water icing after baking, or washed with egg before baking, and left plain.

Cut into 1-inch slices when cool. The fruit may be omitted and many sorts of cookies made out of this mixture.

APPLE SAUCE CAKE

Cream

Brown Sugar	1	lb.
Shortening and Butter	1½	lb.
Salt	⅛	oz.
Cinnamon, Allspice, Mace, and Cloves	¼	oz.

Add

Molasses	12	oz.
Malt Syrup	1½	lb.
Milk	1½	pt.

Sift and add

Bread Flour	2¼	lb.
Cake Flour	2	lb.
Baking Powder	2½	oz.

Add and mix in		
Ground Apples	2	lb.
With Flour	8	oz.
Add		
Currants	1½	lb.
Walnuts	4	oz.
Chopped Cherries	1	lb.
Candied Grapefruit	4	oz.
Lemon and Orange Peel	2	lb.
Seeded Raisins	2	lb.

Bake in loaf cake tin at 350° F.

CHERRY LOAF

Cream		
Sugar, Fine Granulation	1½	lb.
Shortening	1	lb.
Salt	⅛	oz.
Cherry Flavor	¼	oz.
Add		
Egg Whites (Unbeaten)	1	lb.
Add		
Milk	1	lb.
Add		
Maraschino Cherries (Cut up)	1	lb.
Walnuts	¼	lb.
Sift and add		
Cake Flour	2½	oz.
Baking Powder	1½	oz.

Bake in loaf tins at 375° F.

LIGHT FRUIT CAKE

Cream together		
Sugar	2	lb.
Shortening	2	lb.
Salt	1	oz.
Flavor	½	oz.
Then cream in		
Eggs	2	lb.
Next add		
Flour	2	lb.
Baking Powder	¼	oz.

Blend flour and baking powder.

Lastly add		
Almonds	1½	lb.
White Raisins	2	lb.
Citron	2	lb.
Glace Cherries	2	lb.
Glace Pineapple	1½	lb.

Scale into loaf tins and bake at slow heat.

POUND CAKE

Cream together

Sugar	4 lb.
Butter or Shortening	2⅛ lb.
Milk Powder	4 oz.
Flour (Short Patent)	1 lb.

Add

Eggs	2½ lb.
Salt	½ oz.
Flavor	To suit
Invert Syrup	2 oz.

Add

Milk	1 lb.

Add

Flour	3 lb.
Baking Powder	½ oz.

Blend together flour and baking powder.

Then add

Milk	½ lb.

For fruit pound cake, half of the short patent flour should be bread flour. Add about 4–6 pounds of fruit. The fruit should be dusted with about ½ pound bread flour. Bake in wood-lined pound cake tins, 300°–325° F.

GINGER COOKIES

Sugar	2 lb.
Malt Syrup	1 lb.
Shortening	1 lb.
Molasses	1 qt.
Soda	2 oz.
Ginger	1 oz.
Salt	½ oz.
Flour	6 lb.

Cut with 3-inch cutter, and dampen with water. Bake in medium oven.

GINGER SNAPS

Sugar	4 lb.
Malt Syrup	1 lb.
Shortening	30 oz.
Molasses	2 qt.
Water	1 pt.
Soda	2½ oz.
Ammonium Carbonate	½ oz.
Flour	12½ oz.

Cut with 2½-inch cutter and bake in medium oven. These snaps should be washed with an egg wash before baking, or baked in light steam.

HONEY CAKES

Sugar	½ lb.
Shortening	½ lb.
Honey	1 qt.
Malt Syrup	1 lb.
Molasses	1 pt.
Ammonia	1 oz.
Soda	Pinch
Salt	½ oz.
Flour	To make medium dough

Let stand in cool place after mixing, to age for an hour or so, bake in medium oven.

OAT MEAL COOKIES

Sugar	6 lb.
Malt Syrup	2 lb.
Shortening	4 lb.
Eggs	14
Milk	1½ pt.
Soda	3¼ oz.
Raisins (Ground)	2 lb.
Flour	8 lb.
Oats (Rolled)	4 lb.
Salt	1 oz.
Cinnamon	1 oz.

Cut with 2½–3-inch cutter, wash with egg wash, and bake in medium oven.

JELLY ROLL

Sugar	2¼ lb.
Malt Syrup	4 oz.
Shortening	¼ lb.
Egg Yolk	1½ pt.
Milk	1½ pt.
Flour	3 lb.
Baking Powder	1 oz.
Ammonium Carbonate	½ oz.
Salt	½ oz.

This makes a jelly roll or marshmallow roll of good keeping qualities. If a chocolate roll is desired add a paste made of 8 ounces of cocoa and water to make a thick paste.

YELLOW MIX

Sugar	6 lb.
Malt Syrup	1½ lb.
Shortening	3 lb.
Milk	3⅓ qt.
Egg Yolk	1 qt.
Baking Powder	3¼ oz.
Ammonium Carbonate	1 oz.
Flour	11 lb.
Salt	1½ oz.

This is an excellent formula for layers, sheet cakes, and loaf or wine cakes, scale from it. By the addition of a little hard wheat flour and a little more baking powder, jumbles, drop cakes, and a wide variety of small cakes can be run with a pastry bag. The variety is unlimited.

BUTTERMILK COOKIES

Sugar	3½ lb.
Malt Syrup	½ lb.
Shortening	1 lb.
Eggs	15
Soda	3 oz.
Sour or Buttermilk	2 qt.
Flour	7 lb.
Salt	1 oz.
Lemon Flavor	

Drop on greased and dusted pans, bake in hot oven.

DEVIL'S FOOD

Formula No. 1

Sugar, Powdered XXXX	1¾ lb.
Malt Syrup	4 oz.
Shortening	1 lb.
Eggs	6
Sour Milk	3¼ pt.
Milk	¾ pt.
Soda	¾ oz.
Flour	2 lb.
Cocoa	5 oz.

Sift the flour and cocoa together. Can be baked in layers or sheets, and iced with chocolate or marshmallow icing.

DEVIL'S FOOD (Layers or Sheets)

No. 2

Cream together	
Sugar	10 lb.
Shortening	4 lb.
Salt	2 oz.
Soda	2 oz.
Cocoa	1½ lb.
Then add	
Eggs	6 lb.
Vanilla Flavor	To suit
Add mixed together	
Water	2 lb.
Condensed Milk (Sweetened)	4 lb.
Invert Syrup	¾ lb.
Sift together and add	
Flour	9 lb.
Baking Powder	2 oz.
Then add	
Water	2 lb.

Bake at 350°–375° F.

DEVIL'S FOOD (Layers, Cups, and Sheets)

No. 3

Cream together	
Sugar (Fine Granulation)	9 lb.
Shortening	3½ lb.
Salt	2 oz.
Vanilla	2 oz.
Add slowly	
Eggs	4 lb.
Then add	
Milk	4 lb.
Add mixed together	
Flour	10 lb.
Baking Powder	5½ oz.
Add paste of	
Sift Cocoa	2 lb.
Sugar	2 lb.
Soda	¼ lb.
Add	
Cold Water	4 lb.

Bake at 350°–375° F.

CHOCOLATE DOUGHNUTS (PLUNGER MACHINE)

Cream together	
Sugar	6 lb.
Shortening	¾ lb.
Cocoa	1¼ lb.
Salt	3 oz.
Vanilla	2 oz.
Add	
Eggs	3 lb.
Then add	
Milk	10 lb.
Sift together and add	
Bread Flour	5 lb.
Cake Flour	8 lb.
Baking Powder	½ lb.

Fry at 385° F.

DOUGHNUT GLAZE

(1) Granulated Sugar	1 lb.
Agar	3 oz.
Water	1 qt.
Vanilla	If desired

Mix sugar and agar, and boil for 3 minutes in 1 quart of water.

(2) Mix smooth 10 pounds of powdered sugar with 1 quart of hot water.

Pour hot (1) over (2) and mix thoroughly.

FRUIT CAKES

	Golden Fruit Cake	White Fruit Cake	Spice Fruit Cake
Flour	5 pounds	5 pounds	5 pounds
Sugar	3½ pounds	3½ pounds	
Brown Sugar	....	....	5 pounds
Molasses	2 pounds	....	1 pounds
Honey	1 pound	1 pound	
Butter	12 ounces	1 pound	4 ounces
Shortening	3 pounds	3 pounds	3 pounds
Milk	4 ounces	8 ounces	3 pounds
Caramel Color	....	....	3 ounces
Whole Eggs	3 pounds	4 pounds	3 pounds
Salt	3 ounces	2 ounces	4 ounces
Cinnamon	1½ ounces	....	1 ounces
Mace	½ ounce	....	½ ounce
Cloves	1 ounce	....	½ ounce
Vanilla Extract	½ ounce	¼ ounce	¼ ounce
Almond Flavor	....	1 ounce	¼ ounce
Lemon Flavor	1 ounce	....	....
Rum Flavor	1 ounce	....	....
Baking Powder	5 ounce	5 ounce	5 ounce
Fruit Mixture:			
Seedless Raisins	6 pounds	7 pounds	6 pounds
Sliced Citron	3 pounds	3 pounds	1 pound
Sliced Lemon Peel	¼ pound	....	....
Sliced Orange Peel	½ pound	3 pounds	....
Walnut, Pecan Pieces	3 pounds	....	1 pound
Cherry Pieces	2 pounds	3 pounds	2 pounds
Chopped Dates	2 pounds	2 pounds	2 pounds
Water	2 pounds	....	....
Sugar syrup	1 pound	....	....
Almonds	....	2 pounds	....

Cream shortening, add sugar, syrups, beat well, then add egg, milk, etc. Sift flour with baking powder, salt and spices and add to creamed sugar, etc. Sift part of the flour over the fruit, add to batter and add the nuts last. Place in loaf pans lined with greased paper. Cover tightly.

Bake with light steam for 1½–3 hours at 325–350° F.

Remove covers at end of first hour, and finish baking.

Cover fruit mixture with damp cloth overnight.

CHAPTER VII

PREPARED PIE FILLINGS

The manufacturing of pie fillers is very similar to that of making preserves. The essential difference is that starches are used for thickening purposes, instead of pectin.

In making pie fillers insufficient cooking is just as bad for the product as overcooking. In the first instance the product will be too liquid and run out of the pie, while in the second instance the product will be too firm or rubbery, and not pleasant to chew.

It is very important to select the right starches. The manufacturers of starches make starches especially suited for pie fillers. The shortening used in pie fillers should always be of vegetable origin.

APPLE PIE FILLING

Evaporated Apples	50 lb.
Cane Sugar	50 lb.
Corn Syrup	150 lb.
Cornstarch	12 lb.
Tapioca Starch	4 lb.
Mixed Spices	9 oz.
Sodium Benzoate	4 oz.
Salt	8 oz.
Water	10 gal.

1. Soak fruit in water overnight. The following morning soften the fruit with live steam, and then put through the meat grinder.

2. Add 10 gallons of water to kettle, and warm just enough to take chill out of water. Now draw off 3 gallons for the corn starch and 1 gallon for the tapioca, in separate mixing receptacles.

3. To the water left in the kettle add the steamed evaporated apples, start the steam. Then add the cane sugar, mix around to dissolve same, and then add the corn syrup and salt. Now cook the contents of the kettle until the temperature registers around 214° F. At this point add the 4 pounds of tapioca starch, which has been made into a paste with the 1 gallon of water. The batch should be still cooking, not too vigorously, and paddle around the added tapioca until it becomes trans-

parent. Now practically shut off the steam and add slowly the cornstarch, previously made into a paste with the water. Cook up until heavy, and the cornstarch has become transparent.

4. Now add the mixed spices and a solution of the sodium benzoate. Transfer the pie filling to the cooling table, and when cooled down to about 130° F. pour into pails.

BLUEBERRY OR HUCKLEBERRY PIE FILLING

FORMULA No. 1

Fruit	100 lb.
Cane Sugar	75 lb.
Corn Syrup	75 lb.
Sodium Benzoate	4 oz.
Citric Acid	9 oz.
Powdered Pectin No. 80	1½ lb.

1. Mix the pectin powder with about 5 pounds of cane sugar.

2. Add 6 gallons of water to the kettle, bring to a boil, and then add the pectin mixture slowly with mixing. When completely dissolved, add the cane sugar, corn syrup, solution of sodium benzoate, and then the fruit. Bring the contents to a full boil, and when it sheets off the paddle, shut off steam. Now add the solution of citric acid, and mix in. Transfer the filling to the cooling table, and when cool run into pails.

No. 2

Fruit	100 lb.
Cane Sugar	75 lb.
Corn Syrup	75 lb.
Cornstarch	6 lb.
Tapioca Starch	2 lb.
Sodium Benzoate	6 oz.
Citric Acid	9 oz.
Water	10 gal.

1. Add 10 gallons of water to kettle and warm up to remove the chill. Remove 8 gallons to a receptacle and mix in it the 8 pounds of starches.

2. To the 2 gallons left in the kettle add the sugar and the fruit, and start boiling, and while doing so add the corn syrup. Cook to 214° F. and then add the starch paste slowly, with stirring until the starches become transparent. Cook it all up to the proper heaviness. Shut off the steam, and then add the solution of sodium benzoate, stir around. Add the solution of citric acid and stir. Now transfer the filling to the cooling table, paddle around on the table to mix properly, and when cooled down run into pails.

CHERRY PIE FILLING

FORMULA No. 1

Fruit (Montmorency or Morello Cherries)	60	lb.
Cane Sugar	50	lb.
Corn Syrup	45	lb.
Cornstarch	8	lb.
Tapioca Starch	1	lb.
Shortening	3	lb.
Salt	8	oz.
Sodium Benzoate	3	oz.
Cherry Color (Dry)	¼	oz.
Water	6	gal.
Citric Acid	4	oz.

1. Place 6 gallons of water in kettle and warm up to remove the chill. Remove 1½ gallons and ½ gallon into receptacles, for the mixing of the corn and tapioca starch.

2. To the balance of water in the kettle add the sugar, turn on steam, then add the corn syrup, and while cooking add the shortening, salt, and then the fruit. Cook up to 214° F. Now add the tapioca starch paste, and cook to transparency. Then add the cornstarch paste and cook to proper transparency and heaviness. Shut off steam and add a solution of sodium benzoate, then citric acid. Paddle the contents around, transfer to cooling table, and when cooled off pour into pails.

No. 2

Sour Cherries	150 lb.
Cane Sugar	130 lb.
Cornstarch	10 lb.
Tapioca Starch	6 lb.
Citric Acid	8 oz.
Water	12 gal.
Shortening	5 lb.

1. Add 12 gallons of water to kettle and warm up to remove the chill. Now draw off 3 gallons for the cornstarch and 2 gallons for the tapioca, and in separate receptacles make into starch pastes.

2. To the water left in the kettle add the cane sugar, and while cooking add the shortening and then the fruit. Boil the contents up to 214° F. Then add the tapioca starch paste and cook until it becomes transparent. Add the cornstarch paste, cook up to the proper heaviness. Shut off steam and add a solution of citric acid. Now transfer to the cooling table and then into pails.

FIG FILLING

FORMULA No. 1

Fig Paste	50 lb.
Dried Figs	50 lb.
Cane Sugar	100 lb.
Salt	8 oz.
Sodium Benzoate (Optional)	4 oz.
Water	10 gal.

1. Soak the figs and fig paste overnight in water. In the morning put the figs through the meat grinder.

2. Bring the 10 gallons of water in the kettle to a boil and add the cane sugar and then the fig paste. Dissolve the sodium benzoate and salt in a little water and also add to kettle. Bring the contents of the kettle up to 224° F. or to proper heaviness.

No. 2

Fig Paste	50 lb.
Dried Figs	50 lb.
Cane Sugar	100 lb.
Corn Syrup	130 lb.
Salt	8 oz.
Sodium Benzoate (Optional)	5 oz.
Water	10 gal.

Make in same way as recommended above.

No. 3

Dried Figs	100 lb.
Dried Apples	15 lb.
Cane Sugar	100 lb.
Corn Syrup	125 lb.
Sodium Benzoate	5 oz.
Salt	8 oz.
Water	15 gal.

1. Soak apples and figs overnight in separate containers, or the apples may be softened with live steam. Put both through the meat grinder.

2. To the water in the kettle add the cane sugar, start boiling and dissolve the sugar. Add the fig and apple paste, then the corn syrup. Dissolve the salt and sodium benzoate in a little water and add to the kettle. Boil the contents of the kettle to 224° F. or to the proper heaviness.

LEMON PIE FILLING

FORMULA No. 1

Powdered Lemon Juice	¾	lb.
Cane Sugar	110	lb.
Corn Syrup	165	lb.
Shortening	10	lb.
Cornstarch	32	lb.
Tapioca	10	lb.
Powdered Egg Yolk	3	lb.
Citric Acid	1	lb.
Salt	2	lb.
Lemon Oil (Dissolve in alcohol)	5	oz.
Sodium Benzoate	9	oz.
Egg Color Solution	5	oz.
Water	36	gal.

1. Measure out the 36 gallons of water into the kettle, remove chill. Into separate mixing receptacles, draw off 6 gallons for cornstarch, 3 gallons for tapioca starch, ½ gallon for egg powder.

2. Start cooking the balance of the water in the kettle and add the cane sugar, corn syrup, shortening, sodium benzoate, salt, and boil to 213°–214° F.

3. Whip up the tapioca starch with the 3 gallons of water into a paste and add to the kettle, stirring around with the paddle until the starch becomes jelled or transparent. This should take place in a minute or two.

4. Now whip up the cornstarch with the 6 gallons of water, add the color to it, and then add this to the kettle. Cook the batch until the starch jells or becomes transparent. Avoid too long cooking or the resulting pie filler will be too heavy. Paddle the mixture thoroughly after the cornstarch has been added. Shut off steam.

5. Whip up the egg powder until free from lumps with the ½ gallon of water drawn before. Add to kettle and quickly mix it in thoroughly. Now draw off about 3 gallons from the kettle and whip up with the lemon juice powder and also the citric acid dissolved in a little water. Then put it back into the kettle.

Transfer the batch onto the cooling table, and while paddling add the lemon flavor. When cooled down to about 130° F. fill the pails.

EGG COLOR SOLUTION

Tartrazine	9	oz.
Orange I	1	oz.
Water	2½	gal.

LEMON PIE FILLER

	No. 2
Powdered Lemon Juice	4 oz.
Cane Sugar	25 lb.
Corn Syrup	20 lb.
Shortening	2 lb.
Cornstarch	9 lb.
Tapioca	1 lb.
Powdered Egg Yolk	1 lb.
Citric Acid	2 oz.
Salt	2 oz.
Lemon Extract	2 oz.
Sodium Benzoate	2 oz.
Egg Color Solution	½ oz.
Water	52 lb.

Directions: Add to kettle.

Water	37 lb.
Cane Sugar	25 lb.
Corn Syrup	20 lb.
Shortening	2 lb.
Salt	2 oz.

and cook to 214° F.

Water	15 lb.
Tapioca Starch	1 lb.
Cornstarch	9 lb.

Whip up into paste and add with paddling to kettle. Cook until the starches gel, and drop in sheets from the paddle. Shut off the steam.

Whip up with ½ gallon of water the egg powder, lemon powder, citric acid, color, sodium benzoate, and then add to the kettle. Mix around with the paddle, and then transfer to the cooling table. Add flavor to pie filler while on cooling table.

MINCE MEAT PIE FILLING

Water	8 gal.
Evaporated Apples	50 lb.
Brown Sugar	40 lb.
Corn Syrup	100 lb.
Beef Suet	15 lb.
Meat	60 lb.
Seedless Raisins	100 lb.
Spices	1½ lb.
Sodium Benzoate	9 oz.
Salt	1½ oz.
Wine	2 gal.
Alcohol	2 pt.

Soak the apples and raisins overnight. To the water in the kettle add the cane sugar, corn syrup, salt, sodium benzoate, and bring to a boil.

Add the apples and raisins, which have been put through the grinder. Boil the contents to 214° F.

Cook meat until soft, then put through the meat grinder. Transfer to kettle above and cook to proper consistency. Shut off the steam, and now add spices, wine, and alcohol and mix with agitator.

The beef suet after being put through the grinder can be incorporated hot or cold into the batch. When the entire batch is in the cooling pan, or cooling down in the kettle, add the suet and continue agitating, or the beef suet can be added with the meat to the hot batch and thoroughly mixed in.

ORANGE-PINEAPPLE CAKE FILLING

FORMULA No. 1

Crushed Pineapple	48	(No. 10 tins)
Cane Sugar	350	lb.
Corn Syrup	40	lb.
Agar-Agar	2½	lb.
Sodium Benzoate	12	oz.
Salt	2½	lb.
Orange Color *	16	oz.
Orange Oil	8	oz.
Water	8	gal.

1. Soak the agar-agar in 8 gallons of water overnight.

2. Add the pineapple juice to the kettle, bring to a boil. Now add the sugar, corn syrup, salt, a solution of sodium benzoate, and also the color. While still boiling add the fruit and cook up to 216° F. Now add the agar-agar solution, putting it through a strainer first, and continue boiling to 221° F. Shut off the steam, transfer to cooling table, and when temperature is around 140° F. fill into pails.

* ORANGE COLOR SOLUTION

Orange I	4 oz.
Water	1 gal.

ORANGE PINEAPPLE FILLING

No. 2

Crushed Pineapple	24 (No. 10 tins)
Ground Orange Peel	90 lb.
Cane Sugar	400 lb.
Liquid Pectin (See Jelly chapter)	15 gal. (Or 5 gal. Douglas Pectin)
Citric Acid	8 oz.
Water	15 gal.

Add to kettle 15 gallons of water, bring to boil, then add the sugar, pineapple, and ground orange peel. Cook to 221° F. and add the pectin juice. Cook until temperature is 223° F. or until the batch shoots off the paddle. Shut off steam, add a solution of the citric acid, and then transfer to cooking table. A little orange color and sodium benzoate solution may be added, if desired.

PEACH AND APRICOT PIE FILLING

Evaporated Peaches or Apricots	75 lb.
Cane Sugar	75 lb.
Corn Syrup	75 lb.
Sodium Benzoate	5 oz.

Soak peaches or apricots overnight, cook up fruit, sugars, and sodium benzoate, very high, around 224° F.

RASPBERRY PIE FILLING

Raspberry Fruit	25 lb.
Raspberry Pulp	25 lb.
Cane Sugar	10 lb.
Corn Syrup	300 lb.
Red Color	12 oz.
Powdered Pectin No. 80	3½ lb.
Water	15 gal.

1. Mix pectin with the 10 pounds of cane sugar.
2. Boil the 15 gallons of water and then add the pectin powder. When dissolved add the corn syrup, fruit, color, and cook to 221° F., or to a good sheet test.
3. Run hot filling into 30-pound pails, having in them 5 ounces of acid solution. True fruit flavor can be added if wished.

COLOR SOLUTION

Amaranth	4 oz.
Water	1 gal.

ACID SOLUTION

Tartaric Acid	4 lb.
Water	1 gal.

RASPBERRY-PEACH FILLING

This filling is made with the addition of 30 pounds of evaporated peaches to the above formula.

The peaches should be soaked overnight, and in the morning put through the grinder.

BAKERS' STABILIZER C* STRAWBERRY PIE

WITHOUT STARCH

Water	1 gal.
Sugar	5 to 8 lb.
Bakers' Stabilizer C	6 oz.
Salt	2 oz.

Dry mix the stabilizer and sugar. Add to the water. Bring to a boil and boil from 2 to 3 minutes. Remove from fire. Let cool slightly and add frozen strawberries. One pound of strawberries per 1 pound of gel is a proportion frequently used. Variations in the gel may be obtained by using different amounts of stabilizer.

* Stein, Hall & Company, Inc., New York City.

WITH STARCH

Water	1 gal.
Sugar	5 to 8 lb.
Bakers' Stabilizer C	3½ oz.
Salt	2 oz.

Dry mix the stabilizer and sugar. Add to the water. Bring to a boil and boil from 2 to 3 minutes. Add to the boiling solution a suspension of 11 ounces of cornstarch in 1 quart of water. Bring to a boil.

Remove from fire, let cool slightly, and add frozen strawberries. One pound of strawberries per 1 pound of gel is a proportion frequently used. Variations in the gel may be obtained by using different amounts of starch and stabilizer. Thus, when 1 pound of starch is used, the stabilizer should be reduced to 2 ounces.

A few drops of red fruit color may be used in this mixture. The amount of sugar to be used varies, depending on the desired sweetness of the filling.

BAKERS' STABILIZER C* BASE FOR FRUIT PIES

WITHOUT STARCH

Mix 8 ounces of Bakers' Stabilizer C with 10 pounds of sugar and add, with stirring, to 1 gallon of water. Bring to a boil and continue boiling for 2 to 3 minutes. Work in the berries with boiling syrup, allow to cool, and deposit in pie.

The amount of stabilizer necessary varies with different kinds of pies and with different formulas. Each case has to be taken separately. The above is a typical formula with berries, using Bakers' Stabilizer C.

WITH STARCH OR TAPIOCA

Sugar	10 lb.
Water	1 gal.
Bakers' Stabilizer C	3 oz.

Bring to a boil and continue boiling for 2 to 3 minutes. Mix 1½ pounds of cornstarch (or 1 pound of cornstarch and ½ pound of tapioca) with 2 quarts of water. Add to the above and cook until clear. Remove from fire and let cool; then work in the fruit.

* Stein, Hall & Company, Inc., New York City.

CHAPTER VIII

BAKING POWDERS

Powders used as substitutes for yeast in baking are called baking powders. The bread or cake is rendered spongy by the CO_2 (carbonic acid gas) generated in the dough. This is effected by the action of an acid, such as TARTARIC ACID, on SODIUM BICARBONATE and some farinaceous substance is added to act as a diluent.

In making bread add 1 teaspoonful of BAKING POWDER for every pound of flour.

In making cake add 2 teaspoonfuls of BAKING POWDER for every pound of flour.

There are four types of baking powder on the market, commonly known as

I. Cream of Tartar Baking Powder
II. Phosphate Baking Powder
III. Sodium Aluminum Sulfate Baking Powder
IV. Combination Baking Powder

I. CREAM OF TARTAR BAKING POWDERS

Consist of mixtures of cream of tartar ($C_4H_4O_6HK$), molecular weight 188 (potassium hydrogen tartrate); bicarbonate of soda ($NaHCO_3$), molecular weight 84; in proportion of 2 to 1 by weight, and dry starch about ⅕.

In some instances the cream of tartar is partially replaced by tartaric acid ($C_4H_6O_6$), molecular weight 150.

Tartrate powders are expensive but they keep well and are effective when cold. They yield a residue of Rochelle salt (potassium sodium tartrate) ($C_4H_6O_6NaK$), which is soluble in water.

CHEMICAL REACTION

$$KHC_4H_4O_6 + NaHCO_3 \rightarrow NaKC_4H_4O_6 + H_2O + CO_2$$

20% Filler

One level teaspoonful of powder weighs 3 grams and contains 20% starch. This yields approximately 0.4 gram of CO_2 or 200 cubic centimeters at 0° C., which become 273 cubic centimeters at 100° C., the highest temperature of the oven.

The residue of Rochelle salt amounts to 2.5 grams soluble in water.

II. PHOSPHATE BAKING POWDERS

A. CALCIUM PHOSPHATE POWDERS

Consist of soluble acid calcium phosphate ($CaH_4(PO_4)_2$) (mono calcium phosphate); bicarbonate of soda ($NaHCO_3$), and starch as filler.

CHEMICAL REACTION

$$CaH_4(PO_4)_2 + 2NaHCO_3 \rightarrow CaHPO_4 + Na_2HPO_4 + 2CO_2 + 2H_2O$$

Bicarbonate of Soda Acid Calcium Phosphate 25% Filler	Calcium Hydrogen Phosphate	Di Sodium Phosphate	Water

$CaHPO_4$ is insoluble in water. It requires free acid for solution. One level teaspoonful of powder weighs 4.4 grams and contains 25% starch. This yields approximately 0.720 gram CO_2 or 335 cubic centimeters at 0° C., which become 485 cubic centimeters at 100° C., the highest point of the oven.

The residue of phosphates weighs 4.05 grams, 27.5% insoluble in water.

B. SODIUM PHOSPHATE POWDERS

Consist of mono sodium phosphate (NaH_2PO_4), bicarbonate of soda ($NaHCO_3$) and starch.

CHEMICAL REACTION

$$NaH_2PO_4 + NaHCO_3 \rightarrow Na_2HPO_4 + H_2O + CO_2$$

32% Filler

One level teaspoonful of powder weighs 3.75 grams and contains 32% of starch. This yields approximately 0.545 grams CO_2 or 274 cubic centimeters at 0° C., which becomes 374 cubic centimeters at 100° C., the highest point of the oven.

The residue of soluble disodium phosphate weighs 4.41 grams, soluble in water.

Of the phosphate baking powders the SODIUM COMPOUND is undoubtedly the least harmful and the most efficient.

III. ALUM BAKING POWDERS

Consist of sodium aluminum sulphate ($Na_2Al_2(SO_4)_4$) (also called S.A.S.), acid calcium phosphate ($CaH_4(PO_4)_2$), bicarbonate of soda, and starch as filler.

CHEMICAL REACTION (WORKED WITH AMMONIA ALUM)

$$(NH_4)_2Al_2(SO_4)_4 + CaH_4(PO_4)_2 + 4NaHCO_3 \rightarrow$$

$$Al_2(PO_4)_2 + CaSO_4 + (NH_4)_2SO_4 + 2Na_2SO_4 + 4H_2O + 4CO_2$$

33⅓% Filler

The formula is worked with ammonia alum on account of the indefinite composition of S.A.S.

Calcium phosphate and alum phosphate powders are cheap but they do not keep well and leave a residue which is largely insoluble in water.

One level teaspoonful of powder weighs 2.85 grams and contains 33⅓% starch. This yields approximately 0.32 gram CO_2 or 160 cubic centimeters at 0° C., which become 218 cubic centimeters at 100° C. Residue 2.17 grams, 36.6% insoluble in water.

Name	Weight of 1 Teaspoon of Powder	Same Weight Less Filler	Weight of CO_2	Vol. of CO_2 at 0° C.	Vol. of CO_2 at 100° C.	Weight of Residue
	g	g	g	cc	cc	g
Tartrate Baking Powder.....	3	2.4	0.4	200	237	2.5
Ca-Phosphate Baking Powder	4.4	3.3	0.7	355	485	4.05
Na-Phosphate Baking Powder	3.75	2.5	0.5	274	374	4.41
Alum Baking Powder.......	2.85	1.9	0.3	160	218	2..17

TEST FOR CHECKING A BATCH

Mistakes in mixing can be easily caught, independent of the check sheets, by dissolving the powder in pure water and boiling to complete the reaction and testing for marked excess or ALKALI or ACID with litmus paper. An exact mixture gives a purple color, which becomes red in excess of acid and blue in excess of alkali.

DIRECTIONS FOR MIXING A BATCH OF BAKING POWDER

I. Soda and starch, added through the sifter.
Mix for several minutes.

II. Add S.A.S. through sifter.
Let run for 5 minutes.

III. Add phosphates through sifter.
Let run for 18–20 minutes.

BAKING POWDER FORMULAE VERY POPULAR IN THE BAKING TRADE

BAKING POWDER XX

Bicarbonate of Soda	140 lb.
Cornstarch	108 lb.
Sodium Aluminum Sulphate	100 lb.
Acid Calcium Phosphate	52 lb.
Albumen	20 lb.

BAKING POWDER X

The same as XX without the albumen.

BAKING POWDER X2

Take half XX and half X.

BAKING POWDER S

Bicarbonate of Soda	35 lb.
Cornstarch	24 lb.
Sodium Aluminum Sulphate	29 lb.
Acid Calcium Phosphate	12 lb.
Albumen	2 lb.

BAKING POWDER P.M.

Bicarbonate of Soda	36 lb.
Cornstarch	20 lb.
Sodium Aluminum Sulphate	28 lb.
Acid Calcium Phosphate	16 lb.

BAKING POWDER P

Bicarbonate of Soda	102 lb.
Cornstarch	86 lb.
Acid Sodium Phosphate	40 lb.
Acid Calcium Phosphate	45 lb.
Sodium Aluminum Sulphate	72 lb.
Albumen	½ lb.

BAKING POWDER—TYPE C

Sodium Bicarbonate	28 lb.
Cornstarch	41 lb.
Sodium Aluminum Sulphate	19 lb.
Acid Calcium Phosphate	12 lb.

Gas strength: 14½% CO_2

TYPE B

Granulated Sodium Bicarbonate	28 lb.
Redried Cornstarch	37 lb.
Acid Calcium Phosphate	35 lb.

Gas strength: 14½%

MAGIC BAKING POWDER (Like Majestic)

Bicarbonate of Soda	33 lb.
Cornstarch	22 lb.
Mono Sodium Phosphate	45 lb.

CREAM OF TARTAR BAKING POWDER

Bicarbonate of Soda	27 lb.
Cream of Tartar	54 lb.
Cornstarch	14 lb.

BAKING POWDER—SODA SPECIAL

Bicarbonate of Soda	4½ lb.
Tartaric Acid	½ lb.

EGG BAKING POWDER

Cream of Tartar	½ lb.
Tartaric Acid	½ lb.
Bicarbonate of Soda	1½ lb.
Cornstarch	1½ lb.
Powdered Turmeric	1 oz.
Salt	½ oz.

SLOW RISING BAKING POWDER

Cornstarch	40 lb.
Bicarbonate of Soda	27 lb.
Acid Calcium Phosphate	11 lb.
Sodium Aluminum Sulphate	22 lb.

BAKER'S FRIEND BAKING POWDER

Bicarbonate of Soda	36 lb.
Cornstarch	32 lb.
Sodium Aluminum Sulphate	26 lb.
Calcium Acid Phosphate	14 lb.

BAKING POWDER—TARTAR SUBSTITUTE

Acid Sodium Phosphate	40½ lb.
Sodium Aluminum Sulphate	45 lb.
Cornstarch	64½ lb.

Or

Acid Sodium Phosphate	27 lb.
Sodium Aluminum Sulphate	30 lb.
Cornstarch	43 lb.

GROCER'S BAKING POWDER

Bicarbonate of Soda	26 lb.
Acid Calcium Phosphate	33 lb.
Cornstarch	41 lb.

ALUM BAKING POWDER

Bicarbonate of Soda	28 lb.
Sodium Aluminum Sulphate	19 lb.
Acid Sodium Phosphate	20 lb.
Cornstarch	33 lb.

VICTOR'S FORMULAE

BAKING POWDER

Bicarbonate of Soda	35 lb.
Sodium Aluminum Sulphate	25 lb.
Calcium Acid Phosphate	13 lb.
Cornstarch	27 lb.

TARTAR SUBSTITUTE

Bicarbonate of Soda	8 lb.
Sodium Aluminum Sulphate	38 lb.
Acid Calcium Phosphate	27 lb.
Cornstarch	27 lb.

HOUSEHOLD BAKING POWDER

Bicarbonate of Soda	28 lb.
Sodium Aluminum Sulphate	19 lb.
Acid Calcium Phosphate	12 lb.
Starch	41 lb.

GOOD BAKING POWDER

Soda Alum	17 lb.
Calcium Acid Phosphate	20 lb.
Tartaric Acid	2 lb.
Bicarbonate of Soda	28 lb.
Carbonate of Magnesia	2 lb.
Cornstarch	31 lb.

Theoretical Gas: 14¼%

CREAM OF TARTAR BAKING POWDER

Cream of Tartar	61 lb.
Bicarbonate of Soda	28 lb.
Cornstarch	11 lb.

Theoretical Gas: 14½%

ALUM BAKING POWDER

Soda Alum	28 lb.
Bicarbonate of Soda	29 lb.
Cornstarch	43 lb.

Theoretical Gas: 15.4%

PHOSPHATE BAKING POWDER

Granular Calcium Acid Phosphate	56 lb.
Granular Bicarbonate of Soda	25½ lb.
Dried Cornstarch	18½ lb.

Theoretical Gas: 13¼%

CREAM OF TARTAR AND TARTARIC ACID BAKING POWDER

Cream of Tartar	50 lb.
Tartaric Acid	3 lb.
Bicarbonate of Soda	26½ lb.
Cornstarch	20½ lb.

Theoretical Gas: 13¾%

The cream of tartar and the soda should be in the granular form. The starch super dried. Otherwise use more starch.

PHOSPHATE AND ALUM BAKING POWDER

Soda Alum	22 lb.
Calcium Acid Phosphate	11 lb.
Bicarbonate of Soda	27 lb.
Cornstarch	40 lb.

Theoretical Gas: 14¼%

CHEMICALS USED IN BAKING POWDERS

A. ALKALIES

Bicarbonate of soda is in general use. It is easily neutralized, gives off a large amount of gas, is cheap, and is pure commercially.

Besides this, carbonate of magnesia, usually spoken of as "magnesia" ($MgCO_3$), is used in small proportions. It produces about the same amount of gas per pound as $NaHCO_3$, but requires twice the amount of acid to neutralize it. It serves an excellent purpose, being extremely light. Its bulk being about seven times as much as $NaHCO_3$, it increases the bulk of the baking powder. For the same reason, baking powder made with magnesia keeps better because the other ingredients do not lie as closely together. There is another argument in its favor, when used in connection with soda. It is desirable to have enough alkali to neutralize all the excess. To be on the safe side, we want an excess of alkali, but if we have any excess of soda the baking product will be yellow. An excess of magnesia is not objectionable.

B. THE NEUTRAL SUBSTANCES

Flour, cornstarch, rice flour.

Wheat flour often contains larvae which develop worms. Cornstarch being made by a chemical process is the best.

Egg albumen is not of any real economic value. It is used on account of the effect it will produce, in slowing up the powder, when mixed in a glass of water.

Starch and albumen thicken up the water so that a good body of foam is made, while any other baking powder that has only a small amount of starch and no albumen will not foam up in this way.

C. ACIDS

Tartrates {
- Hydric potassium tartrate—Cream of Tartar.
- Dihydric potassium tartrate—Tartaric Acid.

Phosphates

Alums

Cream of tartar is mostly used. Its neutralizing strength is only 45; that is, 100 parts of cream of tartar neutralize but 45 parts of sodium bicarbonate.

When using cream of tartar baking powder, the oven must be very hot, so that the biscuit can be baked immediately, or failure is certain.

The same thing would be true of tartaric acid, but not to the same extent with any other acid ingredient in common use.

Tartaric acid has 2½ times the neutralizing strength of cream of tartar; it is an extremely active acid, and can be used with safety only in small proportions, except if it be in the granular form and with a large proportion of starch. It is a valuable ingredient, however, when used in a proportion of 2 or 3% only, because it begins the leavening process the instant it is wet. Leavening is therefore more continuous and uniform.

Phosphates. The neutralizing strength is about that of cream of tartar. The granular form is believed to possess better keeping qualities.

Acid phosphate of soda (known as pyro sodium phosphate) is thought of very favorably, except for the cost, which is much higher than the phosphate of calcium. A baking powder which contains no other acid except phosphate has never been made with first class keeping qualities; therefore the best way to use it is in connection with alum, whereby its virtues are preserved until they are drawn out by the proper application of moisture and heat.

Alum is a double salt of aluminum and some other base.

Only three of the alums, however, have been in use for baking powder:

Potash }
Ammonia } Alum
Soda }

Potash alum leaves a bitter taste. Ammonia alum is very good, but there is a prejudice against ammonia. Soda alum has no odor in baking and no unpleasant taste. It produces the largest amount of gas and leaves the smallest amount of residue. Soda alum has a neutralizing strength of 97½. It is the cheapest of all acid ingredients, and has the best keeping quality and, as it is most commonly used with phosphates, it preserves all the other ingredients. It takes up all the moisture and thus keeps the baking powder very dry. Alum, when used alone, may be harmful in a small degree.

Chemical action in baking powder occurs when the baking powder comes in contact with moisture and heat. Soda should be used in pure phosphate and pure cream of tartar baking powders, in the granular form. The other ingredients, except the starch, can also be used to advantage in this form, because it has been discovered that the ingredients in the granular form keep much better than if powdered.

CHAPTER IX

MAYONNAISE, SAUCES, CONDIMENTS

Mayonnaise, aside from its popularity as a salad dressing, is also greatly appreciated for its food value. Very often it acts as a base for other salad dressings.

The quality of mayonnaise depends largely upon the freshness of the eggs and the purity of the oil, which are its chief ingredients. A good mayonnaise * or mayonnaise salad dressing is the semi-solid emulsion of edible vegetable oils, egg yolk, or whole egg, a vinegar, and/or lemon juice, with one or more of the following: salt or other seasoning commonly used in its preparation, sugar, and/or dextrose.

MAYONNAISE AND SALAD DRESSINGS

MAYONNAISE

FORMULA NO. 1

High quality. Sharp in taste. Very popular for grocer's trade.

Frozen Egg Yolk	120 lb.
Winter Oil (Cottonseed Oil)	330 lb.
Spice Mixture *	24 lb.
Vinegar **	160 lb.

* Spice Mixture

Salt	200 lb.
Mustard (Ground) No. 1	100 lb.

Mix the frozen egg yolk, which must be taken out of the refrigerator the night before making the batch, with a part of the vinegar and the spices. Be sure to keep some vinegar back to be added last or if the batch breaks. Allow the oil to flow in slowly.

No. 2

Thicker than formula No. 1. Milder in taste. Good keeping qualities. Cheaper.

Frozen Egg Yolk	7½ lb.
Winter Oil (Cottonseed Oil)	33 lb.
Spice Mixture (As in Formula No. 1)	2 lb.
Vinegar (As in Formula No. 1)	10 lb.

* Refer to government definitions and standards for food products.

** Vinegar: Use cider vinegar. Dilute 9 gallons of vinegar with 1 gallon of water.

No. 3

Cottonseed Salad Oil	70.25 lb.
Egg Yolk	10.00 lb.
Vinegar (50 grain)	10.00 lb.
Water	3.90 lb.
Salt	1.45 lb.
Sugar	3.50 lb.
Mustard	.80 lb.
White Pepper	.10 lb.

The egg yolk can be reduced to 7.5 pounds. The sugar can be left out.

No. 4

Cheap. Good keeping qualities. Very popular for restaurants.

Frozen Egg Yolk	4½ lb.
Winter Pressed Cottonseed Oil	75 lb.
Salt	13 oz.
Mustard No. II	8 oz.
White Vinegar	1 qt.
Water	2 qt.

SALAD DRESSING

No. 1

Cornstarch	5 oz.
Water	90 oz.
Cottonseed Oil	51 oz.
French Mustard	25 oz.
Salt	5 oz.
Sugar	10 oz.
Vinegar (White)	42 oz.

Mix the cornstarch with a little cold water and add to the remaining hot water. Stir and boil until it stiffens, set aside to cool. Put the mustard into a bowl and gradually add the oil, stirring all the time. Next add sugar, salt, vinegar, and then the cooled-down cornstarch.

No. 2

Tapioca Starch	7 oz.
Water	30 oz.
Corn or Cottonseed Oil	40 oz.
Salt	2 oz.
Cane Sugar	4 oz.
Vinegar (White 5%)	17 oz.

SAUCES AND OTHER CONDIMENTS

FRENCH DRESSING

Into

Water	1 gal.
Cider Vinegar	8 gal.
Worcestershire Sauce	2 qt.

Mix by stirring or agitation

Mustard Flour	2	lb.
Paprika	5	lb.
Salt	13	lb. 13 oz.
Black Pepper	9⅗	oz.
Sugar	20	lb.

until the salt and sugar are dissolved, which may require 5–10 minutes. Then add

Olive Oil	8	gal.
Cottonseed Oil	2	gal. 2¾ qt.

and agitate vigorously until there is an emulsion formed before filling bottles.

MINT SAUCE

Granulated Sugar	120 lb.
Vinegar	16 gal.
Water	16 gal.
Spearmint Oil	4 fl. oz.

Sugar and water must be mixed and agitated previous to the addition of the vinegar and spearmint. Final batch must stand overnight before bottling.

TOMATO CATSUP

Bruised Tomatoes	1	gal.
Salt	½	lb.

Let stand for 3 or 4 days. Squeeze out the juice and to every ½ gallon add

Black Pepper	1	oz.

Boil ½ hour, strain and add

Allspice	½	oz.
Ginger	½	oz.
Mace	½	oz.
Cochineal	¼	oz.

Simmer gently ½ hour; strain, cool, and bottle. A little brandy may be added, and shallots, if the flavor is liked, may be boiled along with the black pepper.

TOMATO CATSUP (Cheap)

Tomato Pulp	20	gal.
Vinegar	2	gal.
Salt	4½	pt.
Powdered Cloves	½	lb.
Powdered Allspice	½	lb.
Powdered Black Pepper	¼	lb.
Powdered Red Pepper	1	oz.

APPLE CHUTNEY

Apple chutney is prepared from the fresh apples, peeled, cored, and cut into pieces about half an inch cube. The exact shape of the pieces is not important so long as they are not too small. The apples, after chopping, are allowed to stand overnight and then drained from any juice that may have separated, the latter being reserved.

To every 60 pounds of apples 100 pounds of sugar is weighed out, made into a syrup with water, and boiled to 240° F. Into this syrup the small quantity of juice that may have separated is incorporated. While still boiling hot the syrup is poured onto the chopped apples in a suitable container, stirred and allowed to stand for 24 hours. The syrup and apple is then placed in a pan and boiled gently.

together with chopped raisins, chopped stem ginger, and as much spice (such as mace, pimento, and nutmeg) and vinegar as taste demands, and the product bottled hot. Served with cold meat—particularly ham and pork—and similar dishes, this chutney is delightful. The color should be golden brown, but this can be darkened if desired with a little sugar caramel. The only machinery required, apart from the boiling pan, is a chopping or dicing machine.

SALTED NUTS

Use high-grade cocoanut oil.

This oil must be frequently changed or it will produce salted nuts which become stale very rapidly.

Heat the oil to 300° F. When up to temperature drop in basket with nuts. When thoroughly cooked, the basket is raised out of the oil, the hot nuts are emptied into a cooling pan, and then the salt is spread on at that time. It is necessary to use something to help hold the salt on the nuts, glucose, gum arabic, or some prepared product.

Use 4 pounds of salt to 100 pounds of nuts.

PEANUT BUTTER

Use

Spanish Peanuts	50 lb.	Freshly roasted
and		
Virginia Peanuts	50 lb.	

Pass twice through a fine meat grinder. The last time add 2 pounds of salt before grinding.

MIXED PICKLES

To 1 gallon PICKLING VINEGAR (see this) add

Cauliflower	20–30%
Small Onions	10%
Cucumbers	60–70%

and other vegetables like tomatoes, tabasco peppers, and tie the jar well. The pickles will be ready in about 3 months.

HORSERADISH SAUCE

Grated Horseradish	1 lb.
Salt	1 oz.
Sugar	2 oz.
Vinegar	1 qt.

Pour the vinegar, boiling hot, over the others, and let stand for a week. Then bottle.

DISTILLED VINEGAR, IMITATION *

Distilled Water	25 gal.
Acetic Acid (80%)	½ gal.
Acetic Ether	2 oz.
Caramel	4 oz.
Sugar Syrup	4 oz.
Acidity about 2%	1 qt.

* See Government specifications of various vinegars.

CIDER VINEGAR, IMITATION*

Place in a keg 1 gallon of good sweet cider, add 20 gallons of water and ½ gallon of acetic acid (80%). Let stand 2 weeks and add 10 more gallons of water and let stand 2 more weeks and use. It gets stronger with age, so that more water may be added from time to time.

PICKLING VINEGAR

To each gallon of vinegar add

Bruised Ginger	3 oz.
Mustard	3 oz.
Salt	4 oz.
Mustard Seeds	2 oz.
Turmeric	1 oz.
Red Pepper	1 dr.
Laurel Leaves	1 oz.

Let stand for 1 month, mix every morning, then strain.

MALT VINEGAR

Made by fermenting barley malt; is generally used for pickling. Strength 45 grains.**

CIDER VINEGAR

Has 4½–5% of natural acetic acid and traces of malic acid.

WHITE WINE VINEGAR

Is the best. It has a little more acetic acid than cider vinegar and traces of tartaric acid.

CELERY SALT

To 5 pounds of salt add 1 fluid ounce of celery seed oil. Darken with 1 pound of powdered celery seeds.

BROWN GRAVY SALT

Salt	8 oz.
Brown Sugar	4 oz.
Cayenne Pepper, Powdered	10 gr.

Mix.

PREPARED CHICKEN STUFFING

To

Poultry Seasoning	3¼ lb.

Add

Thyme	¾ lb.
and	
Black Pepper	¾ lb.

To this mixture add

Whole Wheat Cracker Meal	4–227 grams

* See Government specifications of various vinegars.

** Grain. To say that a vinegar is "40 grain" is the same as saying that it is 4% acetic acid strength.

SAUSAGE SEASONING

Sage	½ lb.
Pepper	1 lb.
Allspice	1 lb.
Salt	1 lb.
Or	
Cayenne Pepper or Capsicum	1 oz.
Cumin	1 oz.
Cassia	1 oz.
Nutmeg	2 oz.
Pimento (Allspice)	6 oz.
Black Pepper	8 oz.
Salt	8 oz.

SOUP SPICE

Lemon Peel	½ lb.
Thyme	1 lb.
Sweet Marjoram	½ lb.
Summer Savory	½ lb.
Parsley, Dried	1 lb.
Celery Seed	¼ lb.
Sweet Brazil	¼ lb.

CURRY POWDERS

Coriander	300 lb.
White Pepper	30 lb.
Red Pepper*	30 lb.
Yellow Mustard	30 lb.
African Ginger	15 lb.
Mace	7½ lb.
Cumin Seed	60 lb.
Fennel Seed	7½ lb.
Cardamom	30 lb.
Turmeric	75 lb.
Foenugreek	30 lb.
Saigon Cassia	25 lb.
Mint	18 lb.
Celery Seed	7½ lb.
Cloves	7½ lb.
Or	
Coriander	3 lb.
Black Pepper	1 lb.
Cayenne Pepper	¼ lb.
Mustard (Yellow)	1 lb.
Jamaica Ginger	1 lb.
Cumin Seed	¼ lb.
Cardamom	½ lb.
Turmeric	3 lb.
Cloves	¼ lb.
Allspices	½ lb.

* If a milder curry is desired, leave out the red pepper.

HOT CURRY

Coriander	1½ lb.
Cumin	½ lb.
Turmeric	1 lb.
Ginger	2 oz.
Mustard	1 oz.
Foenugreek	1 oz.
Cayenne Pepper	1½ oz.

TABLE SAUCE

Allspice	2 oz.
Black Pepper	1 oz.
Cloves	1 oz.
Mustard	8 oz.
Brown Sugar	3 lb.
Salt	1 lb.
Vinegar	4 gal.

Mix and place on the fire and simmer for about 2 hours. Then strain through a cloth and add enough sugar color to give it the desired dark shade. Let stand for 2 weeks before using. Keep in a keg; it will improve with age.

CHILI SAUCE

Ripe Tomatoes, Chopped Fine	5 lb.
Vinegar	2 pt.
Garlic	1 oz.
Red Pepper	1 dr.
Salt	2 oz.
Lemon Juice	5 oz.

Boil for an hour, then force through a colandor and bottle while warm. Cork tight.

TABLE MUSTARD

Mustard Flour	1 lb.
Vinegar	1 qt.
Jamaica Pepper	1 dr.

Boil the vinegar and pour over the mustard. Stir until all lumps are gone, add the pepper and let stand several days, well covered. Add salt to suit. Force through a collander, if necessary.

FRENCH MUSTARD

The same as table mustard with the addition of 8 ounces of sugar and a trace of powdered cloves.

GERMAN MUSTARD

Laurel Leaves	4 oz.
Cinnamon	½ oz.
Cardamom	2 dr.
Sugar	8 oz.
Vinegar	1 qt.

Heat in a double boiler for 1 hour, strain, and add enough mustard flour to make a thin paste. Let stand 1 week.

ENGLISH MUSTARD

Made by mixing mustard flour with vinegar and a little salt.

BUTCHER'S PEPPER

Is coarse ground black pepper.

PEPPERETTE

Ground olive stones, used to dilute pepper.

LONG PEPPER

Long, nearly cylindrical spikes, covered with closely packed fruit which are picked unripe.

The Java long pepper consists of spikes about 4 to 6 centimeters in length, the other long peppers are about half as long and are more common. Long pepper contains less piperin than most black pepper and has a disagreeable odor and flavor.

In the ground state it is not a recognized article of commerce.

It is used whole in pickling and has been used as an adulterant in ground black and white pepper.

To detect long pepper spread the sample of ground pepper, with a spatula, in a smooth, thin layer; pieces of fluffy, woody fibers will be present if long pepper has been used.

BOTTLED RELISHES

OLIVES

Put-up olives shall be of the kind and quality known to the trade as "Prime Queens" and shall count 40–45 to a 16-ounce bottle.

STUFFED OLIVES

Stuffed olives shall be the "Seville Fina Manzanilla" olives, pitted, and stuffed with Spanish pimentos, and shall count 85–95 olives to a 16-ounce bottle. There shall be no broken olives.

PICKLED ONIONS

The onions shall be from ½–¾ inch in diameter, packed in pure white grain vinegar with spices.

CHOW-CHOW PICKLES

Chow-chow mixture shall contain:

60–70% pickles, part whole, part cut. The whole pickles shall not be larger than 150 to a gallon, free from nubs and soft stock.
20–30% cauliflower, cut uniform in size.
10% onions, small size, larger ones to be cut in two, smaller ones left whole.
10% slices of green tomatoes from sound stock, free from yellow or partly ripe tomatoes and not exceeding 1½ inches in diameter.
1 or several small tabasco peppers may be used, and sufficient chow dressing to freely cover the pickles and fill to two-thirds of neck of bottles.

DILL PICKLES

Dill pickles shall be prepared from good, green cucumbers, thoroughly cured in brine. The pickles to be crisp and of good flavor, 4–5 inches in length. Rough stock, badly curved, nubs and seedy stock to be rejected. They shall be free from all preservatives except salt and aluminum sulphate.

10-gal. kegs	280 pickles to the keg
30-gal. casks	900 pickles to the cask

PICKLED GHERKINS

Gherkins are small cucumbers with a prickly skin. Pickled gherkins shall be prepared from good, clean, sound cucumbers, first thoroughly cured in brine and subsequently in vinegar. The pickles shall be crisp and of good flavor. Rough stock, nubs, badly curved and seedy stock to be rejected. They shall be uniform in size, measuring not less than 2 or more than 2½ inches in length, and free from all preservatives except salt, vinegar, sugar, spices, and aluminum sulphate.*

MIXED PICKLES

The mixture shall contain **

60–70% pickles, part whole, part cut. The whole pickles to be not larger than 150 to a gallon, free from crooks, nubs, and soft stock.

20–30% cauliflower, cut uniform in size, no core to be used.

10% onions, small size, larger ones to be cut in two, smaller ones left whole.

10% sliced green tomatoes, from sound stock, free from yellow or partly ripe tomatoes, and not exceeding 1½ inches in diameter.

1 or several small tabasco peppers may be used and sufficient vinegar to fully cover the pickles and fill to two-thirds of neck of bottle.

MANGO PICKLES

Pickled mango melons or pickled green peppers stuffed with finely chopped pickles.
"Nubbins," "Crooks," and "Nubs"—small, imperfectly formed cucumbers.

PICKLES

Drained Weights

50 lb. of drained pickles is the weight established for a bushel of thoroughly cured salt stock.

47½ lb. of drained pickles is the weight established for a bushel of pickles shipped within 3 months after salting.

VINEGARS

Vinegar is the acid liquid resulting from the acetous fermentation of various decoctions or fruit juices. Acetic acid is the prominent constituent, but small amounts of alcohol, aldehyde, and ester (ethyl acetate) are usually present, together with extractive matters, depending upon the nature of the material used. Very dilute solutions of acetic acid do not keep well, and the presence of a little alcohol is regarded as desirable, improving the flavor and keeping qualities. Some mineral acid was formerly thought to be necessary as a preservative. Such addition is not needed, and is now sometimes practiced as an adulteration.

Vinegar is often made by spontaneous fermentation, but malt and spirit vinegars are mostly made by passing a dilute alcohol over beech shavings or birch twigs, impregnated with the acetic ferment, principally mycoderma aceti, a regulated supply of air being maintained at the same time. The conversion of the alcohol into acetic acid takes place rapidly. Wine, cider, malt, and spirit vinegar are the chief forms.

* Aluminum sulphate is added to effect the greening of the pickles. This is allowed. But the addition of cupric sulphate is against the law, as it is very poisonous. To detect cupric sulphate, incinerate the pickles, treat the ash with a little nitric acid, and add an excess of ammonia to the solution. In the presence of copper, a blue coloration will be produced.

** These specifications are taken from the Department of Purchases of the City of New York.

All vinegars made by fermentation without distillation must carry in solution only the extractive matter derived exclusively from the substances from which they were fermented. All vinegars which contain any added drugs, acids, coloring matter of any kind, or ingredients not derived exclusively from the substances from which they were respectively made, shall be deemed adulterated. Also if they contain less than 4% acetic acid.

YEAST

Alcoholic fermentation depends upon the development of a minute fungoid growth, named MYCODERMA CEREVISIAE or SACCHAROMYCES VINI.

This constitutes principally the brewer's yeast, of which there are two varieties, the upper or top yeast (Oberhefe of the Germans) and the lower or bottom yeast (Unterhefe of the Germans).

The former is the kind most generally met with in the United States. It is a viscid, semifluid, frothy mass, containing large, more or less globular, microscopic cells from which smaller ones are developed. It appears upon the surface of the fermenting liquid (a mixture of corn, rye, barley, malt, and water or diluted molasses of a density of 1.060) and is reproduced therein at a temperature ranging between 15°–20° C. But a fermenting liquid (wort) fermented with top yeast at a temperature near or below 10° C. will produce both top and bottom yeast. At a temperature of about 9° C. and less only bottom yeast will be reproduced. It consists of various sized cells, differing from those of the top yeast in not having any small cells attached and appears to be reproduced from isolated spores, while the top yeast multiplies by budding. Compressed yeast is marketed in small cakes. Dry yeast is yeast mixed with cornstarch and then dried. It is less active than the compressed but will keep for a longer time.

Fermentation is divided into two distinct classes:

1. Organized ferments, such as yeast, bacteria, etc.
2. Unorganized ferments or enzymes.

The first are called organized ferments because they contain life. The second are chemical substances which are not living organisms but are produced by such.

ENZYMES

INVERTASE	Changes cane sugar into invert sugar.
ZYMASE	Acting on sugars, produces alcohol and carbon dioxide.
DIASTASE	Changes raw starch into soluble starch and then into maltose and dextrose.
MALTASE	Changes dextrin and maltose into dextrose.
CYTASE	Changes cellulose into sugars.
PECTASE	Changes pectin into pectates and sugar.
LACTASE	Changes lactose into dextrose and galactose.

HOW TO PREPARE BRINE

BRINE FOR KRAUT

Use 5 ounces of salt for each gallon of water.

BRINE FOR SOUR PICKLES

Use diluted distilled vinegar in the proportion of 1 gallon of water to each 4 gallons of vinegar.

BRINE FOR SWEET PICKLES

Use vinegar in the proportion of 1 gallon of water to each 4 gallons of vinegar and about 3 to 4 pounds of granulated sugar for each gallon of liquor.

BRINE FOR DILL PICKLES

Use 8 ounces of salt to 1 gallon of water.

BRINE FOR SALT FISH

Use 2 pounds of coarse salt to every gallon of water.

BRINE FOR SPICED FISH

Use 1 quart of pickling vinegar to every 3 quarts of water.

BRINE FOR GREEN OLIVES

Use 12 ounces of good salt and ½ ounce of 50% lactic acid to each gallon of water. In summer use 14 ounces of salt.

Do not use metal utensils in the preparation of any brine.

CHAPTER X

ANALYSES AND TESTS

TESTS AND ANALYSES OF EDIBLE OILS AND FATS
ACTION OF NITRIC ACID ON DIFFERENT OILS

Use concentrated nitric acid having a specific gravity of 1.42. Shake equal parts of oil and acid in a large test tube and note change of color.

Oil	First Action	After Five Minutes
Sweet Almond	Flesh tint	Cherry red
Lard	Lemon or yellow to brown	Chocolate red to brown red
Linseed, Raw	Olive green, then dark brown	Brown to reddish brown
Olive	Greenish	Dirty brown
Cottonseed	Brown	Chocolate brown to deep blood red
Rape	Pale rose	Orange yellow
Nut	Pale rose	Orange yellow
Sesame	White	Brown yellow
Arachis (Peanut)	Pale rose	Brown yellow
Corn	Red	

ELAIDIN TEST

One cubic centimeter of mercury is dissolved in 12 cubic centimeters of nitric acid. Two cubic centimeters of this freshly made deep green solution are shaken in a wide-mouthed stoppered bottle with 50 cubic centimeters of the sample to be tested and the agitation repeated every 10 minutes.

A HARD MASS IS YIELDED BY

Olive oil, almond oil, lard oil, arachis oil.

A PRODUCT OF THE CONSISTENCY OF BUTTER

Mustard oil, and sometimes arachis and rape oil.

A PASTY MASS WHICH SEPARATES FROM A FLUID PORTION

Rape oil, sesame oil, cottonseed oil, sunflower oil, and sometimes mustard oil.

LIQUID BY-PRODUCTS

Linseed oil, hempseed oil, walnut oil, and other drying oils.

The results of the Elaidin test must be accepted with caution.

ELAIDIN TEST

Oil	Coloration	Minutes for Solidification
Olive	Pale yellowish green	60
Almond	White	90
Arachis	Pale reddish	105
Rape	Orange	200
Cottonseed	Orange red	105
Sesame	Yellowish orange	150
Beechnut	Reddish orange	360
Poppy	Red	remains fluid

SETTIMI TEST

FOR SOY OIL AND OLIVE OIL

Shake vigorously so as to form a thorough emulsion

Oil	5 cc.
Chloroform	2 cc.
2% Aqueous Uranium Nitrate Solution	3 cc.
Intense Yellow Color	Soy Oil
White Color	Peanut, Sesame, Corn Oil
Green Color	Olive Oil

HALPHEN TEST

FOR COTTONSEED OIL

Mix carbon disulphide (CS_2) containing about 1% sulphur in solution with an equal volume of amyl alcohol. Mix equal volumes of this reagent and the oil sample and heat in a bath of boiling saturated brine for 1 to 2 hours. In the presence of as little as 1% of cottonseed oil, a characteristic red or orange red color is produced.

PHYTOSTEROL TEST

FOR CORN OIL

The oil is dissolved in carbon disulphide, a drop of strong sulphuric acid added, and the liquid allowed to stand for 24 hours. A fine violet coloration will be produced with corn oil.

FURFURAL TEST

FOR SESAME OIL (OIL OF BENNE)

Dissolve 0.1 gram of sugar in 10 cubic centimeters of concentrated hydrochloric acid. Add 20 cubic centimeters of sample. Shake 1 minute. Let stand. Aqueous layer becomes crimson if sesame oil is present.

TEST FOR FISH AND VEGETABLE OILS

Dissolve in a test tube about 6 grams of sample in 12 cubic centimeters of a mixture of equal parts of chloroform and glacial acetic acid.

Add bromine, drop by drop, until slight excess is indicated by color. Keep at 20° C.

After 15 minutes place tube in boiling water.

IF VEGETABLE OIL	IF FISH OIL
Solution perfectly clear.	Cloudy or precipitate.

CHEMICAL ANALYSIS OF OILS

SPECIFIC GRAVITY

Use a clean and perfectly dry pycnometer. Weigh this empty, then with distilled water. Empty the pycnometer and dry it with alcohol and ether. Fill with the oil to be tested and place in a bath of constant temperature until the oil reaches the temperature at which the specific gravity is desired to be taken. Weigh and calculate by dividing the weight of oil by the weight of the distilled water. That is the specific gravity.

EXAMPLE

68.4625 g. = weight of pycnometer and water (temperature 15° C.)
18.4035 g. = weight of pycnometer

50.0590 g. = weight of distilled water

64.1620 g. = weight of pycnometer and oil (temperature 15° C.)
18.4035 g. = weight of pycnometer

45.7585 g. = weight of oil
45.7585 : 50.0590 = 0.914 = specific gravity of oil at 15° C.

IODINE NUMBER (HANUS METHOD)

Weigh 0.250 gram of the oil into a 500-cubic centimeter glass-stoppered flask. Add 10 cubic centimeters of chloroform to dissolve the oil. Use part of the chloroform to wash out the weighing bottle in which the 0.250 gram was weighed out. Dissolve the oil in the chloroform by shaking.

Add 25 cubic centimeters of Hanus' iodine solution (can be purchased in any chemical supply house). Allow to stand 30 minutes. Shake occasionally. Then add 10 cubic centimeters of 15% potassium iodide solution. Shake well.

Now add 100 cubic centimeters of distilled water and titrate with N/10 sodium thiosulphate solution, using starch solution as the indicator.

Make a blank determination, similar in every detail, except that no oil is added. Deduct the number of cubic centimeters of iodine used by the blank from the number of cubic centimeters absorbed by the oil. The difference gives the true number of cubic centimeters of iodine absorbed by the oil. Multiply by 1.269 and divide by the weight of oil taken.

EXAMPLE

17.472 g. = weight of oil and weighing bottle
17.222 g. = weight of weighing bottle

0.250 g. = weight of oil

Used for titration of the oil = 34.3 cc. N/10 sodium thiosulphate solution.
Used for titration of the blank = 53.7 cc. N/10 sodium thiosulphate solution.

53.7
−34.3

19.4

Factor for N/10 sodium thiosulphate solution = 1.0093

19.4 × 1.0093 = 19.58042
19.58042 × 1.269 = 24.84702
24.84702 : .250 = 99.4

Therefore 99.4 is the iodine number of the oil.

ANALYSIS OF VINEGAR

TEST FOR CIDER VINEGAR

To 10 cubic centimeters of the vinegar add a few drops of neutral lead acetate solution.

If the sample is cider vinegar a precipitate will form.

TEST FOR MINERAL ACIDS (LIKE SULPHURIC ACID)

Five-hundredths gram of pulverized starch is dissolved in 10 cubic centimeters of the vinegar by boiling, and after the liquid has become completely cooled a few drops of iodine solution are added. Dilute acetic acid (vinegar) does not affect starch, and if the sample is pure a blue coloration will be produced. If, however, but a minute quantity of sulphuric acid or other mineral acid is present, the starch is converted into dextrine and the addition of iodine fails to cause the blue coloration.

CHEMICAL ANALYSIS

TOTAL ACIDS

Dilute 10 cubic centimeters of the sample with distilled water until it appears very slightly colored (about 30–50 cubic centimeters) and titrate with 0.5/N alkali (sodium hydroxide or potassium hydroxide).

Use phenolphthalein solution as an indicator.

1 cc. 0.5/N alkali = 0.030 g. CH_3COOH (acetic acid)

TOTAL SOLIDS (RESIDUE)

Evaporate 50 cubic centimeters of the sample.

The solids obtained upon evaporation to dryness should approximate 2%.

ANALYSES AND TESTS FOR FRESH EGGS, DRIED EGGS, EGG ALBUMEN AND EGG YOLK

CANDLING EGGS

A very convenient form of candling device can be made by any tinsmith. The device should have the form of a cylinder-shaped chimney similar to the glass chimneys used on lamps. A small opening should be made about one-half to three-fourths of the way on the side of the cylinder, to which should be soldered a cone-shaped piece of tin having another opening on the bottom. The opening should be just large enough to allow the egg to pass through, put lengthwise. By placing this apparatus over the top of a small oil lamp, it is possible to pass the egg through the opening on the side of the cone-shaped chamber and observe it through the light. If candling is carried on in a dark room, the baker can with very little practice determine quite accurately the condition of the egg.

In order to become thoroughly familiar with candling eggs, it is advisable to select a series of from 8 to 10 eggs, ranging from an egg 1 to 2 days old up to eggs 10 weeks old. By examining a series of this kind, it will be noticed that in the fresh egg the yolk will be very much intact and the air chamber on the upper portion of the egg will be very small. As progression is made in the series, it will be observed that the yolk is less intact and that the air chamber gradually increases in size. After having examined a series of this kind it will be possible with very little practice to approximately determine the age of the eggs under examination.

EXAMINING THE EGGS ON A PORTION OF THE PLATE

After the series of eggs have been examined by candling it is advisable to break each egg carefully onto a plate and arrange them in order. By carefully observing the eggs on a portion of the plate it will be found that there will be a difference in the symmetry and the appearance of the yolk as well as the white. The yolk of the fresh egg will be as before, very much intact, while the white will have a firm formation. As progress is made in this examination, it will again be observed that the yolk gradually loses its firmness while the white becomes more watery. Very often in making examinations of eggs it will be found that the egg will have blood streaks and little clots known as meat spots. Meat spots may be found in fresh eggs, while streaks are the result of incubation.

FRESH EGGS

GRADE SPECIFICATIONS

For Nearby Fancy Grade, Grade A and Grade B, ten or more eggs in each dozen must meet the standard for the grade as given below. Two eggs in each dozen may be below the standard specification, but may not be below the standard for the nest lower grade. For Grade C and Unclassified Grade each lot of eggs offered for sale must meet the requirements as given.

Grade Factors	Nearby Fancy Grade	Grade A	Grade B	Grade C	Unclassified Grade
Size	Uniform	Uniform	Uniform	May be variable	May be variable
Shell	Clean, sound	Clean, sound	Clean, sound	May be clean or dirty, sound or cracked	May be clean or dirty, sound or cracked.
Air-Cell (Candling Test)	1/8″ or less in depth, localized, regular.	2/8″ or less in depth, localized, regular	3/8″ or less in depth, localized, may be slightly tremulous	May be over 3/8″ in depth, may be bubbly or freely mobile	No fixed standards for this grade. Eggs may be variable in quality, but none may be below the requirements for C
White (Candling)	Firm, clear	Firm, clear	Reasonably firm	May be weak and watery	As above
Germ (Candling)	No visible development	No visible development	Development may be slightly visible	Development but no blood	May be clearly visible showing

FRESH EGGS

The term "fresh" may be applied, in addition to the proper grade designation, to any lot of eggs sold as a unit, which are free from objectionable odor and flavor and which meet the requirements for Nearby Fancy or Grade A(not below requirements of Grade A).

The specific gravity of fresh eggs is about 1.090. After 30 days it may be 1.035.

A simple method of determining the specific gravity and inferring the age consists in dissolving 2 ounces of salt in 1 pint of water and noting the depth to which the eggs sink in this solution.

An egg 1 day old will sink below the surface, but not to the bottom, while one 3 days old will swim just immersed in the liquid. If more than 3 days old, the egg will float on the surface, the amount of shell exposed increasing with age. If 2 weeks old, only a little of the shell will dip in the liquid.

ANOTHER METHOD TO DETERMINE THE AGE OF EGGS

Place the egg in a 10% solution of salt.

Perfectly fresh eggs sink to the bottom. Those remaining immersed, but suspended in the liquid, are at least 3 days old. Those rising to the surface and floating are more than 5 days old. The older the egg, the higher it floats and the more it will stand on end.

Not applicable to preserved eggs.

DETECTING COLD STORAGE EGGS

By dipping eggs in lamp black, one can tell immediately whether they are freshly laid or cold storage.

The test depends upon the fact that storage eggs are treated with an oil to preserve them. If it is a cold storage egg, the lamp black will cling readily to the outer shell, while the amount of lamp black adhering to a fresh egg is negligible.

DRIED EGGS

Moisture	4.61– 8.5 %
Protein	40.4 –49.7 %
Lecithin and Fat	31.7 –45.12%
Ash	3.42– 4.27%
	4.7 – 7.43%
	1.18– 1.37%

A moisture and ash determination is all the compounder can do in his laboratory. The results are best compared with a known sample of a good quality.

ALBUMEN OR EGG WHITE

Egg albumen is generally transparent and of a light yellow color.

On treatment with cold water commercial albumen of good quality should dissolve almost completely. In making the test, the albumen should always be added to the water and not vice versa.

Five grams of albumen added to 50 cubic centimeters of water and leaving no residue is considered a high-grade sample.

MOISTURE DETERMINATION

Dry to constant weight at 100° C. in an oven, cool and weigh.

ASH DETERMINATION

Treat the residue from the moisture determination with a few drops of nitric acid C.P. (HNO_3) and 2–3 drops of sulphuric acid C.P. (H_2SO_4). Heat gently and evaporate the yellow liquid to dryness. Ignite, dry and weigh. Report as "Sulphate Ash."

Egg Albumen	7.0–7.4%
Serum Albumen	6.2–9.7%

Addition of sugar or dextrin lowers the ash. If abnormally high, test for inorganic metals as zinc.

ALBUMEN WHIPPING TEST

This test is quickly made and shows whipping quality.

Place 6 ounces of water in the whipping bowl and add 1 ounce of albumen. Whip until stiff. (For comparison tests one should whip all samples a given time and then compare the volume obtained.) Leave the whipped albumen overnight in a funnel with long stem, close the stem with a cork. Note the amount of water that leaked out during the night.

Commercial albumen is frequently adulterated with dextrins, gums, sugar, flour, etc.

Add 5 grams slowly to 50 cubic centimeters of water until the soluble matter is dissolved. (Pure and high-grade samples leave no residue.)

Add a few drops of acetic acid and filter through silk or fine muslin into a 500-cubic centimeter flask.

The insoluble residue may consist of coagulated albumen, casein, starch, etc.

Treat the albumen with very dilute caustic soda solution, filter and then exactly neutralize the filtrate with acetic acid. If casein is present it will be dissolved by the caustic soda and precipitated on neutralization.

Make the original filtrate up to 500 cubic centimeters. Pour about 100 cubic centimeters into a beaker and heat to boiling. This should coagulate the albumen. Filter and treat the filtrate with a little acetic acid and potassium ferrocyanide solution to make sure that no proteins remain in the solution.

If there is a precipitate, filter it out (a precipitate generally indicates casein, although it must be remembered that any zinc would be thrown out as white ferrocyanide).

Cool the filtrate and add a little concentrated tannin solution. This will precipitate any gelatin or glue. Filter and concentrate the filtrate to a small bulk. Cool and treat with a considerable excess of alcohol. Any precipitate indicates the presence of gums or dextrins. Filter, boil off all the alcohol, treat with diluted HCl and test the solution in the usual way with Fehling's solution, after neutralizing the excess hydrochloric acid (HCl). Any reduction indicates, but does not necessarily prove, the presence of sugar. Sugar may also be extracted by treating the original solid sample with alcohol.

SOLUBLE COAGULABLE ALBUMEN

Pipette 100 cubic centimeters of the original aqueous solution (equivalent to 1 gram) into a beaker. Add about 1 gram of sodium acetate and heat to boiling. Filter the flocculent precipitate on a tared filter, using a platinum cone or Gooch crucible and suction. Wash with hot water. Dry at 100° C. and weigh.

SPRAY EGG YOLK

To 1 ounce of yolk add enough water to obtain a moist paste. Then add 3 more ounces, a little at a time, stir, and let stand for 3 hours. No in-between layer should be formed.

DETERMINATION OF TOTAL SOLIDS, MOISTURE, AND EGG WHITES IN FROZEN EGG YOLK

Two microscopical glass slides, perfectly clean and dry, are weighed on an analytical balance. One is supplied with a small amount (say 0.2 gram) of the yolk, spread evenly on it with a spatula. The other slide is placed over it and both are weighed again. Now place the slides, taken apart, in an electric vacuum drying oven and dry for 1–2 hours at 75° C. and 25 pounds pressure (vacuum). Let cool in desiccator; weigh. Figure from the data obtained the amount of total solids and look in the "Table" for equivalent per cent of egg whites.

(Based on the formula $W = 133.1 - 2.68S$)
$W = \%$ Egg Whites
$S = \%$ Total Solids

EXAMPLE

10.630 grams = weight of glass slide and egg yolk (moist)
10.143 grams = weight of glass slide
.487 grams = weight of yolk (moist)

Dried, cooled, and re-weighed

10.380 grams = weight of glass slide and yolk (dry)
10.143 grams = weight of glass slide
.237 grams = weight of yolk (dry)

CALCULATION

$.487 : .237 = 100 : X$
$X = 48.6$

The sample has 48.6% total solids.
In the Table 48.6% total solids = 51.4% Moisture
= 2.9% Egg Whites

ANALYSIS OF GELATIN

MOISTURE

Weigh out 3 grams dry at 110° C. to constant weight (16–19% moisture is normal).

ASH

The dried sample is transferred to a weighed platinum crucible. Burn, starting with a small flame (1–3% ash is normal). Above 4% the addition of chemicals is suspected.

TEST FOR ALUM

Extract ash with hot water. Filter. Add ammonium chloride solution (NH_4Cl) and enough ammonia (NH_4OH) to obtain strong ammonia odor. A white flocculent precipitate indicates the presence of alum.

SOLUBILITY, CLEARNESS, REACTION, ODOR, JELLY STRENGTH, TASTE

Weigh 4 grams of gelatin into a beaker. Add 96 cubic centimeters of water (that makes a 4% solution). Allow the gelatin to soak in the cold water for ½ hour. Then put the glass into a bath or double boiler (water bath) and turn on the heat. By constant stirring the gelatin begins to dissolve.

Test with litmus paper for reaction (alkaline, acid, or neutral).

If neutral or acid	Bone glue
If alkaline	Hide glue

TABLE

PER CENT OF EGG WHITE CORRESPONDING TO TOTAL SOLIDS IN COMMERCIAL EGG YOLKS

Analysis			Analysis		
Per Cent Total Solids	Per Cent Moisture	Per Cent Egg White	Per Cent Total Solids	Per Cent Moisture	Per Cent Egg White
49.7	50.3	0.0	44.8	55.2	13.1
49.6	50.4	0.2	44.7	55.3	13.3
49.5	50.5	0.5	44.6	55.4	13.8
49.4	50.6	0.9	44.5	55.5	13.9
49.3	50.7	1.1	44.4	55.6	14.1
49.2	50.8	1.2	44.3	55.7	14.3
49.1	50.9	1.3	44.2	55.8	14.7
49.0	51.0	1.9	44.1	55.9	14.9
48.9	51.1	2.1	44.0	56.0	15.1
48.8	51.2	2.2	43.9	56.1	15.3
48.7	51.3	2.5	43.8	56.2	15.9
48.6	51.4	2.9	43.7	56.3	16.0
48.5	51.5	3.1	43.6	56.4	16.2
48.4	51.6	3.3	43.5	56.5	16.6
48.3	51.7	3.8	43.4	56.6	16.9
48.2	51.8	4.1	43.3	56.7	17.1
48.1	51.9	4.2	43.2	56.8	17.3
48.0	52.0	4.4	43.1	56.9	17.8
47.9	52.1	4.8	43.0	57.0	17.9
47.8	52.2	5.0	42.9	57.1	18.1
47.7	52.3	5.2	42.8	57.2	18.3
47.6	52.4	5.3	42.7	57.3	18.7
47.5	52.5	5.9	42.6	57.4	18.9
47.4	52.6	6.1	42.5	57.5	19.1
47.3	52.7	6.3	42.4	57.6	19.3
47.2	52.8	6.5	42.3	57.7	19.9
47.1	52.9	7.0	42.2	57.8	20.1
47.0	53.0	7.1	42.1	57.9	20.3
46.9	53.1	7.3	42.0	58.0	20.5
46.8	53.2	7.7	41.9	58.1	20.8
46.7	53.3	7.9	41.8	58.2	21.0
46.6	53.4	8.1	41.7	58.3	21.3
46.5	53.5	8.3	41.6	58.4	21.7
46.4	53.6	8.9	41.5	58.5	21.9
46.3	53.7	9.1	41.4	58.6	22.1
46.2	53.8	9.2	41.3	58.7	22.3
46.1	53.9	9.5	41.2	58.8	22.9
46.0	54.0	9.9	41.1	58.9	23.0
45.9	54.1	10.1	41.0	59.0	23.2
45.8	54.2	10.2	40.9	59.1	23.5
45.7	54.3	10.8	40.8	59.2	23.9
45.6	54.4	10.9	40.7	59.3	24.0
45.5	54.5	11.1	40.6	59.4	24.2
45.4	54.6	11.3	40.5	59.5	24.5
45.3	54.7	11.8	40.4	59.6	24.9
45.2	54.8	12.0	40.3	59.7	25.1
45.1	54.9	12.3	40.2	59.8	25.3
45.0	55.0	12.5	40.1	59.9	25.8
44.9	55.1	12.8	40.0	60.0	25.9

NOTE THE ODOR!

When the solution is clear and no traces of particles of gelatin can be seen, put the beaker into a refrigerator and leave there overnight. The stronger gelatin will solidify first.

A good gelatin will not melt in this test below 28°–30° C. The higher the melting temperature, the better the grade of the gelatin.

The jelly will be set the next morning when the beaker is taken out of the refrigerator into a well-lighted room for inspection. Note the color and appearance. Should be a clear, transparent jelly, light in color. Taste the jelly. Then with the cold finger press down on the jelly. The degree of resistance to pressure will give the strength. SEE SPECIFICATION OF PLAIN GELATIN.

One can become quite an expert with a little practice in this test. A real expert can tell the difference to a fifth of a grade, and that is as close as the most complicated mechanical devices can come. An instrument called the "Penetrator" can be used, if so desired, and can be purchased in any chemical wholesale house that deals in chemical apparatus.

MELTING AND SETTING TEST

High-melting and quick-setting gelatin is the best. Make a 4% gelatin solution (4 grams of gelatin and 96 cubic centimeters of water) and melt on the water bath. A good gelatin will not melt below 28°–30° C. An inferior grade will melt at 25° C. The same solution on cooling should completely set at 20°–23° C.; an inferior grade sets, often only at 17°–18° C.

MESH (If the gelatin is used for "Dessert Powder")

The ground gelatin shall be 30–50 mesh. Sift the gelatin through a 20-inch sieve, weigh what passes through. Sift the remainder through a 40-inch sieve, weigh what passes through. Sift the remainder through a 60-inch sieve, weigh what passes through.

BACTERIA TEST

For a simple yet conclusive bacteria test, allow the beaker with the 4% gelatin solution to remain covered up in the laboratory and examine twice a day. In holding the glasses between you and the light, you will see in many gelatins little spots in the jelly. They are various forms of bacteria developing. In some gelatins these bacteria colonies can be detected in less than 24 hours. This indicates a gelatin containing MILLIONS TO THE GRAM.

As you examine the glasses from time to time, you will note that the growth of colonies has a marked effect on the solidity of the jelly. By shaking the glass you observe that the jelly is rapidly turning to liquid, which indicates that the strength of the gelatin is being undermined by liquefying bacteria. A gelatin may go 36 hours before showing bacteria growth, but if a sample can be kept at room temperature for several days without signs of colonies, you have a gelatin that is practically sterile.

SOUNDNESS

Allow the 4% gelatin solution, after cooking, to set uncovered at a temperature of 65°–75° F. for 24 hours. No sign of decomposition at any point in the jellied mass must be noticed. No liquefying bacteria, no mold, no odor, no bad taste.

SPECIFICATIONS FOR PLAIN GELATIN (NOT FLAVORED)

QUALITY

Gelatin must be edible, clean, pure, and a sound product free from offensive odor, granulated, of clear and light color, and of high jelly strength.

JELLY STRENGTH

A proportion of 400 cubic centimeters of water added to ½ ounce of gelatin (3½% solution) must make a firm enough jelly to permit the turning of the glass receptacle containing the gelatin at an angle of 45 degrees for 1 minute without the gelatin breaking away from the side of the container.

SOUNDNESS

The gelatin shall show no sign of decomposition at any point in the jellied mass, 24 hours after cooking the solution; same being allowed to set uncovered at a temperature of 65° to 75° F. during this time.

PURITY

ARSENIC	Not over 1.4 parts in a million
SULPHUR DIOXIDE	Not over 350 parts in a million
LEAD	None
COPPER	Not over 30 parts in a million
ZINC	Not over 100 parts in a million

SPECIFICATIONS FOR GELATIN DESSERT POWDER

Gelatin must be of 30 to 50 mesh. It must yield a firm jelly at a temperature of 60° F. (15.5° C.). Sugar contents not less than 80% best quality pulverized sugar.

COPY OF A GELATIN TEST REPORT

SAMPLE SHIPMENT FROM John Doe, New York
MARKS 1 W
DATE March 2, 1934

MESH	Through 20 2%
	Through 40 98%
(For dessert gelatin only)	Through 60 None
	Through 30–50 98%
COLOR	White, light straw
MOISTURE	12%
ASH	3%
SOLUBILITY	O.K.
(Clearness)	Clear
REACTION	Slightly acid
ODOR	
COLD	After soaking: Good
HOT	After setting: Sl. glue-like
JELLY STRENGTH	95 (100 taken as standard)
TASTE	Little sharp
MELTING TEST	30° C.
SETTING TEST	20° C.
BACTERIA TEST	After 3 days: No clouds, very faint odor
	After 5 days: No clouds, slight odor
	After 12 days: Slight mold
SOUNDNESS	
(After 24 hours)	Perfect
REMARKS	Passed for Contract Department

(Signed) N. N.,
Chemist

ANALYSIS AND TESTS FOR SUGARS

TEST FOR SUCROSE (Cane Sugar, Beet Sugar)

Sucrose will not reduce Fehling's solution. Prepare a 5% solution of cobaltous nitrate and a 50% solution of potassium hydroxide (KOH). If the sugar to be tested is in solution and contains dextrin or gums, remove these at first with alcohol. Fifteen cubic centimeters of the sugar solution are mixed with 5 cubic centimeters of the cobaltous nitrate reagent and 2 cubic centimeters of KOH solution are added.

Sucrose, permanent amethyst blue. Dextrose (glucose), first turquoise blue, passing into light green. In a mixture of the two sugars the color due to sucrose will predominate.

TEST FOR GLUCOSE

Glucose will reduce Fehling's solution.

TROMMER TEST

To the solution to be tested add 1 to 2 drops of copper sulphate solution, then 4 to 5 cubic centimeters of sodium carbonate solution. Heat gently. If glucose is present the blue color will disappear and a reddish brown to yellow precipitate will be formed.

PICRIC ACID TEST

A solution of glucose is heated to 90° C. with a small amount of soda solution and mixed with a solution of picric acid, boiling finally. A blood red color will appear.

CHEMICAL ANALYSIS

MINERAL MATTER AND IMPERFECT REFINEMENT

Put 25 grams into a 100-cubic centimeter graduated measure, shake occasionally till all the sugar is dissolved. If well refined, colorless or but slightly cloudy. If imperfect, yellow brown or nearly black. Let remain undisturbed for 12 hours for mineral matter to deposit, and then read off the per cent.

MOISTURE

Weigh off 5 grams in a dried weighing dish, place in oven at 140° F. (60° C.), and let dry for 12 hours, then weigh again. Calculate the loss in weight as water.

ASH

Take 5 grams of the same, moisten it with water, and then with pure sulphuric acid (H_2SO_4). Heat this gently until frothing ceases, and continue to burn to redness. Take off gas or out of the heated chamber, and moisten the sugar a second time with H_2SO_4. Let it burn again until the ash is quite white. Then cool down and carefully weigh. High-class sugars, when dissolved in water, should not change blue litmus paper to red. (Free acid.) Add a few drops of weak acid solution to some of the solid sugar in a test tube and notice the unpleasant smell of sulphurated hydrogen in the presence of blue ultramarine, which might be added to whiten the sugar. The presence of this is confirmed by bringing a filter paper, moistened with silver nitrate solution, over the mouth of the tube. A dark, almost black, color indicates the presence of sulphides, which are constituents of ultramarine. Instead of silver nitrate solution, lead acetate (sugar of lead) solution may be used.

TEST FOR IRON SALTS

1. Place some sugar solution in a test tube, add a few drops of weak nitric acid and some solution of ferrocyanide of potassium. A greenish-blue coloration and precipitate indicates iron salts.

2. Place some sugar solution in a test tube, add a few drops of weak nitric acid and some solution of sulphocyanide of potassium. A pale pink to blood red coloration indicates iron salts.

3. Place some sugar solution in a test tube, add a few drops of tannic acid solution. A light-colored precipitate gradually darkening to black indicates iron salts.

ANALYSIS FOR COMMERCIAL GLUCOSE (Corn Syrup)

Glucose of good quality should give a clear solution in water. It should contain no unconverted starch, which would cause a turbidity. It must not redden blue litmus paper (free acid). No unpleasant after-taste.

CANDY TEST

To determine the suitability of liquid glucose for hard-boiled candy (3% water or less), it is necessary to be sure that it is of the type which will produce less than 8.5% of inversion.

Test: Boil

Glucose	1 lb.
Sugar	2 lb.
Water	4½ lb.

in copper pan to 320° F. in 25 minutes. Cool. Determine dextrose in a 5% solution.

QUANTITATIVE ANALYSIS OF SUGAR (SUCROSE)

A. INVERSION OF THE SUGAR

(Conversion of the sucrose into dextrose and levulose.)

Weigh 20 grams of sugar into a 100-cubic centimeter graduated flask and dissolve with 75 cubic centimeters of water. Bring the solution up to 100-cubic centimeter mark with water.

Fifty cubic centimeters of this solution are placed in a flask, 5 cubic centimeters of hydrochloric acid added, and the liquids well mixed.

The flask is heated in water until the thermometer registers 68° C. About 15 minutes should be required for this heating. The flask is then cooled and the contents neutralized with caustic soda solution.

B. DETERMINATION OF SUGAR AS DEXTROSE

Transfer 30 cubic centimeters of Fehling's solution A and 30 cubic centimeters of Fehling's solution B and 50 cubic centimeters of the above-mentioned sugar solution to a 400-cubic-centimeter beaker. Heat this solution over a Bunsen flame, regulating the heat so that the solution begins to boil in 4 minutes, and continue the boiling for exactly 2 minutes. Keep the beaker covered with a watch glass during the heating. Filter the solution at once, wash with water, then with 10 cubic centimeters of alcohol, and finally with 10 cubic centimeters of ether. Dry the residue on a weighed filter, cool in a desiccator, and weigh as Cu_2O (cuprous oxide).

Find the per cent of dextrose in Allihn's Table for the determination of dextrose.

ALLIHN'S TABLE[1] FOR THE DETERMINATION OF DEXTROSE

(Expressed in Milligrams.)

Copper	Cuprous Oxide	Dextrose	Copper	Cuprous Oxide	Dextrose	Copper	Cuprous Oxide	Dextrose
12	13.5	7.1	102	114.8	51.9	192	216.2	98.4
14	15.8	8.1	104	117.1	52.9	194	218.4	99.4
16	18.0	9.0	106	119.3	54.0	196	220.7	100.5
18	20.3	10.0	108	121.6	55.0	198	222.9	101.5
20	22.5	11.0	110	123.8	56.0	200	225.2	102.6
22	24.8	12.0	112	126.1	57.0	202	227.4	103.7
24	27.0	13.0	114	128.3	58.0	204	229.7	104.7
26	29.3	14.0	116	130.6	59.1	206	231.9	105.8
28	31.5	15.0	118	132.8	60.1	208	234.2	106.8
30	33.8	16.0	120	135.1	61.1	210	236.4	107.9
32	36.0	17.0	122	137.4	62.1	212	238.7	109.0
34	38.3	18.0	124	139.6	63.1	214	240.9	110.0
36	40.5	18.9	126	141.9	64.2	216	243.2	111.1
38	42.8	19.9	128	144.1	65.2	218	245.4	112.1
40	45.0	20.9	130	146.4	66.2	220	247.7	113.2
42	47.3	21.9	132	148.6	67.2	222	249.9	114.3
44	49.5	22.9	134	150.9	68.2	224	252.4	115.3
46	51.8	23.9	136	153.1	69.3	226	254.4	116.4
48	54.0	24.9	138	155.4	70.3	228	256.7	117.4
50	56.3	25.9	140	157.6	71.3	230	258.9	118.5
52	58.5	26.9	142	159.9	72.3	232	261.2	119.6
54	60.8	27.9	144	162.1	73.4	234	263.4	120.7
56	63.0	28.8	146	164.4	74.4	236	265.7	121.7
58	65.3	29.8	148	166.6	75.5	238	268.0	122.8
60	67.6	30.8	150	168.9	76.5	240	270.2	123.9
62	69.8	31.8	152	171.1	77.5	242	272.5	125.0
64	72.1	32.8	154	173.4	78.6	244	274.7	126.0
66	74.3	33.8	156	175.6	79.6	246	277.0	127.1
68	76.6	34.8	158	177.9	80.7	248	279.2	128.1
70	78.8	35.8	160	180.1	81.7	250	281.5	129.2
72	81.1	36.8	162	182.4	82.6	252	283.7	130.3
74	83.3	37.8	164	184.6	83.7	254	286.0	131.4
76	85.6	38.8	166	186.9	84.8	256	288.2	132.4
78	87.8	39.8	168	189.1	85.9	258	290.5	133.5
80	90.1	40.8	170	191.4	86.9	260	292.7	134.6
82	92.3	41.8	172	193.6	87.9	262	295.0	135.7
84	94.6	42.8	174	195.9	89.0	264	297.2	136.8
86	96.8	43.9	176	198.1	90.0	266	299.5	137.8
88	99.1	44.9	178	200.4	91.1	268	301.7	138.9
90	101.3	45.9	180	202.6	92.1	270	304.0	140.0
92	103.6	46.9	182	204.9	93.1	272	306.2	141.1
94	105.8	47.9	184	207.1	94.2	274	308.5	142.2
96	108.1	48.9	186	209.4	95.2	276	310.7	143.3
98	110.3	49.9	188	211.7	96.3	278	313.0	144.4
100	112.6	50.9	190	213.9	97.3	280	315.2	145.5

[1] Z. Ver. Ruebenzucker-Ind., **32** (N. F. 19), 606, 865 (1882)

ALLIHN'S TABLE—*Continued*

(Expressed in Milligrams)

Copper	Cuprous Oxide	Dextrose	Copper	Cuprous Oxide	Dextrose	Copper	Cuprous Oxide	Dextrose
282	317.5	146.6	342	385.0	179.8	402	452.6	214.1
284	319.7	147.7	344	387.3	180.9	404	454.8	215.2
286	322.0	148.8	346	389.6	182.1	406	457.1	216.4
288	324.2	149.9	348	391.8	183.2	408	459.4	217.5
290	326.5	151.0	350	394.0	184.3	410	461.6	218.7
292	328.7	152.1	352	396.3	185.4	412	463.8	219.9
294	331.0	153.2	354	398.6	186.6	414	466.1	221.0
296	333.3	154.3	356	400.8	187.7	416	468.4	222.2
298	335.5	155.4	358	403.1	188.9	418	470.6	223.3
300	337.8	156.5	360	405.3	190.0	420	472.9	224.5
302	340.0	157.6	362	407.6	191.1	422	475.6	225.7
304	342.3	158.7	364	409.8	192.3	424	477.4	226.9
306	344.5	159.8	366	412.1	193.4	426	479.6	228.0
308	346.8	160.9	368	414.3	194.6	428	481.9	229.2
310	349.0	162.0	370	416.6	195.7	430	484.1	230.4
312	351.3	163.1	372	418.8	196.8	432	486.4	231.6
314	353.5	164.2	374	421.1	198.0	434	488.6	232.8
316	355.8	165.3	376	423.3	199.1	436	490.9	233.9
318	358.0	166.4	378	425.6	200.3	438	493.1	235.1
320	360.3	167.5	380	427.8	201.4	440	495.4	236.3
322	362.5	168.6	382	430.1	202.5	442	497.6	237.5
324	364.8	169.7	384	432.3	203.7	444	499.9	238.7
326	367.0	170.9	386	434.6	204.8	446	502.1	239.8
328	369.3	172.0	388	436.8	206.0	448	504.4	241.0
330	371.5	173.1	390	439.1	207.1	450	506.6	242.2
332	373.8	174.2	392	441.3	208.3	452	508.9	243.4
334	376.0	175.3	394	443.6	209.4	454	511.1	244.6
336	378.3	176.5	396	445.9	210.6	456	513.4	245.7
338	380.5	177.6	398	448.1	211.7	458	515.6	246.9
340	382.8	178.7	400	450.3	212.9	460	517.9	248.1
						462	520.1	249.3

MUNSON AND WALKER'S TABLE[1] FOR CALCULATING DEXTROSE, INVERT SUGAR ALONE, INVERT SUGAR IN THE PRESENCE OF SUCROSE (0.4 GRAM AND 2 GRAMS TOTAL SUGAR), LACTOSE, LACTOSE AND SUCROSE (2 MIXTURES), AND MALTOSE (CRYSTALLIZED).

(Expressed in Milligrams.)

Cuprous Oxide (Cu_2O)	Copper (Cu)	Dextrose (*d*-Glucose)	Invert Sugar	Invert Sugar and Sucrose		Lactose	Lactose and Sucrose		Maltose	Cuprous Oxide (Cu_2O)
				0.4 gram total sugar	2 grams total sugar	$C_{12}H_{22}O_{11}+H_2O$	1 lactose, 4 sucrose	1 lactose, 12 sucrose	$C_{12}H_{22}O_{11}+H_2O$	
10	8.9	4.0	4.5	1.6	....	6.3	6.1	....	6.2	10
12	10.7	4.9	5.4	2.5	....	7.5	7.3	....	7.9	12
14	12.4	5.7	6.3	3.4	....	8.8	8.5	....	9.5	14
16	14.2	6.6	7.2	4.3	....	10.0	9.7	....	11.2	16
18	16.0	7.5	8.1	5.2	....	11.3	10.9	....	12.9	18
20	17.8	8.3	8.9	6.1	....	12.5	12.1	....	14.6	20
22	19.5	9.2	9.8	7.0	....	13.8	13.3	....	16.2	22
24	21.3	10.0	10.7	7.9	....	15.0	14.5	....	17.9	24
26	23.1	10.9	11.6	8.8	....	16.3	15.8	....	19.6	26
28	24.9	11.8	12.5	9.7	....	17.6	17.0	....	21.2	28
30	26.6	12.6	13.4	10.7	4.3	18.8	18.2	....	22.9	30
32	28.4	13.5	14.3	11.6	5.2	20.1	19.4	....	24.6	32
34	30.2	14.3	15.2	12.5	6.1	21.4	20.7	....	26.2	34
36	32.0	15.2	16.1	13.4	7.0	22.8	22.0	....	27.9	36
38	33.8	16.1	16.9	14.3	7.9	24.2	23.3	....	29.6	38
40	35.5	16.9	17.8	15.2	8.8	25.5	24.7	....	31.3	40
42	37.3	17.8	18.7	10.1	9.7	20.6	26.0	....	32.9	42
44	39.1	18.7	19.6	17.0	10.7	28.3	27.3	....	34.6	44
46	40.9	19.6	20.5	17.9	11.6	29.6	28.6	....	36.3	46
48	42.6	20.4	21.4	18.8	12.5	31.0	30.0	....	37.9	48
50	44.4	21.3	22.3	19.7	13.4	32.3	31.3	....	39.6	50
52	46.2	22.2	23.2	20.7	14.3	33.7	32.6	....	41.3	52
54	48.0	23.0	24.1	21.6	15.2	35.1	34.0	....	42.9	54
56	49.7	23.9	25.0	22.5	16.2	36.4	35.3	....	44.6	56
58	51.5	24.8	25.9	23.4	17.1	37.8	36.6	....	46.3	58
60	53.3	25.6	26.8	24.3	18.0	39.2	37.9	....	48.0	60
62	55.1	26.5	27.7	25.2	18.9	40.5	39.3	....	49.6	62
64	56.8	27.4	28.6	26.2	19.8	41.9	40.6	....	51.3	64
66	58.6	28.3	29.5	27.1	20.8	43.3	41.9	....	53.0	66
68	60.4	29.2	30.4	28.0	21.7	44.7	43.3	40.7	54.6	68
70	62.2	30.0	31.3	28.9	22.6	46.0	44.6	41.9	56.3	70
72	64.0	30.9	32.3	29.8	23.5	47.4	45.9	43.1	58.0	72
74	65.7	31.8	33.2	30.8	24.5	48.8	47.3	44.2	59.6	74
76	67.5	32.7	34.1	31.7	25.4	50.1	48.6	45.4	61.3	76
78	69.3	33.6	35.0	32.6	26.3	51.5	49.9	46.6	63.0	78
80	71.1	34.4	35.9	33.5	27.3	52.9	51.3	47.8	64.6	80
82	72.8	35.3	36.8	34.5	28.2	54.2	52.6	49.0	66.3	82
84	74.6	36.2	37.7	35.4	29.1	55.6	53.9	50.1	68.0	84
86	76.4	37.1	38.6	36.3	30.0	57.0	55.3	51.3	69.7	86
88	78.2	38.0	39.5	37.2	31.0	58.4	56.6	52.5	71.3	88

[1] U. S. Bur. Standards Circ. 44, p. 321. The columns headed "Lactose" and "Lactose and Sucrose" are taken from "Methods of Sugar Analysis and Allied Determinations" by Arthur Given.

MUNSON AND WALKER'S TABLE—*Continued*

(EXPRESSED IN MILLIGRAMS.)

Cuprous Oxide (Cu_2O)	Copper (Cu)	Dextrose (*d*-Glucose)	Invert Sugar	Invert Sugar and Sucrose		Lactose	Lactose and Sucrose		Maltose	Cuprous Oxide (Cu_2O)
				0.4 gram total sugar	2 grams total sugar	$C_{12}H_{22}O_{11}$ $+H_2O$	1 lactose, 4 sucrose	1 lactose, 12 sucrose	$C_{12}H_{22}O_{11}$ $+H_2O$	
90	79.9	38.9	40.4	38.2	31.9	59.7	57.9	53.7	73.0	90
92	81.7	39.8	41.4	39.1	32.8	61.1	59.3	54.9	74.7	92
94	83.5	40.6	42.3	40.0	33.8	62.5	60.6	56.0	76.3	94
96	85.3	41.5	43.2	41.0	34.7	63.8	61.9	57.2	78.0	96
98	87.1	42.4	44.1	41.9	35.6	65.2	63.3	58.4	79.7	98
100	88.8	43.3	45.0	42.8	36.6	66.6	64.6	59.6	81.3	100
102	90.6	44.2	46.0	43.8	37.5	68.0	66.0	60.8	83.0	102
104	92.4	45.1	46.9	44.7	38.5	69.3	67.3	62.0	84.7	104
106	94.2	46.0	47.8	45.6	39.4	70.7	68.6	63.2	86.3	106
108	95.9	46.9	48.7	46.6	40.3	72.1	70.0	64.4	88.0	108
110	97.7	47.8	49.6	47.5	41.3	73.5	71.3	65.6	89.7	110
112	99.5	48.7	50.6	48.4	42.2	74.8	72.6	66.7	91.3	112
114	101.3	49.6	51.5	49.4	43.2	76.2	74.0	67.9	93.0	114
116	103.0	50.5	52.4	50.3	44.1	77.6	75.3	69.1	94.7	116
118	104.8	51.4	53.3	51.2	45.0	79.0	76.7	70.3	96.4	118
120	106.6	52.3	54.3	52.2	46.0	80.3	78.0	71.5	98.0	120
122	108.4	53.2	55.2	53.1	46.9	81.7	79.3	72.7	99.7	122
124	110.1	54.1	56.1	54.1	47.9	83.1	80.7	73.9	101.4	124
126	111.9	55.0	57.0	55.0	48.8	84.5	82.0	75.1	103.0	126
128	113.7	55.9	58.0	55.9	49.8	85.8	83.4	76.3	104.7	128
130	115.5	56.8	58.9	56.9	50.7	87.2	84.7	77.5	106.4	130
132	117.3	57.7	59.8	57.8	51.7	88.6	86.0	78.7	108.0	132
134	119.0	58.6	60.8	58.8	52.6	90.0	87.4	79.7	109.7	134
136	120.8	59.5	61.7	59.7	53.6	91.3	88.7	81.1	111.4	136
138	122.6	60.4	62.6	60.7	54.5	92.7	90.1	82.3	113.0	138
140	124.4	61.3	63.6	61.6	55.5	94.1	91.4	83.5	114.7	140
142	126.1	62.2	64.5	62.6	56.4	95.5	92.8	84.7	116.4	142
144	127.9	63.1	65.4	63.5	57.4	96.8	94.1	85.9	118.0	144
146	129.7	64.0	66.4	64.5	58.3	98.2	95.4	87.1	119.7	146
148	131.5	65.0	67.3	65.4	59.3	99.6	96.8	88.3	121.4	148
150	133.2	65.9	68.3	66.4	60.2	101.0	98.1	89.5	123.0	150
152	135.0	66.8	69.2	67.3	61.2	102.3	99.5	90.8	124.7	152
154	136.8	67.7	70.1	68.3	62.1	103.7	100.8	92.0	126.4	154
156	138.6	68.6	71.1	69.2	63.1	105.1	102.2	93.2	128.0	156
158	140.3	69.5	72.0	70.2	64.1	106.5	103.5	94.4	129.7	158
160	142.1	70.4	73.0	71.2	65.0	107.9	104.8	95.6	131.4	160
162	143.9	71.4	73.9	72.1	66.0	109.2	106.2	96.8	133.0	162
164	145.7	72.3	74.9	73.1	66.9	110.6	107.5	98.0	134.7	164
166	147.5	73.2	75.8	74.0	67.9	112.0	108.9	99.2	136.4	166
168	149.2	74.1	76.8	75.0	68.9	113.4	110.2	100.4	138.0	168

MUNSON AND WALKER'S TABLE—*Continued*

(Expressed in Milligrams)

Cuprous Oxide (Cu_2O)	Copper (Cu)	Dextrose (*d*-Glucose)	Invert Sugar	Invert Sugar and Sucrose		Lactose	Lactose and Sucrose		Maltose	Cuprous Oxide (Cu_2O)
				0.4 gram total sugar	2 grams total sugar	$C_{12}H_{22}O_{11}$ $+H_2O$	1 lactose, 4 sucrose	1 lactose, 12 sucrose	$C_{12}H_{22}O_{11}$ $+H_2O$	
170	151.0	75.1	77.7	76.0	69.8	114.8	111.6	101.6	139.7	170
172	152.8	76.0	78.7	76.9	70.8	116.1	112.9	102.8	141.4	172
174	154.6	76.9	79.6	77.9	71.7	117.5	114.3	104.1	143.0	174
176	156.3	77.8	80.6	78.8	72.7	118.9	115.6	105.3	144.7	176
178	158.1	78.8	81.5	79.8	73.7	120.3	117.0	106.5	146.4	178
180	159.9	79.7	82.5	80.8	74.6	121.6	118.3	107.7	148.0	180
182	161.7	80.6	83.4	81.7	75.6	123.1	119.7	108.9	149.7	182
184	163.4	81.5	84.4	82.7	76.6	124.3	121.0	110.1	151.4	184
186	165.2	82.5	85.3	83.7	77.6	125.8	122.4	111.3	153.0	186
188	167.0	83.4	86.3	84.6	78.5	127.2	123.7	112.5	154.7	188
190	168.8	84.3	87.2	85.6	79.5	128.5	125.1	113.8	156.4	190
192	170.5	85.3	88.2	86.6	80.5	129.9	126.4	115.0	158.0	192
194	172.3	86.2	89.2	87.6	81.4	131.3	127.8	116.2	159.7	194
196	174.1	87.1	90.1	88.5	82.4	132.7	129.2	117.4	161.4	196
198	175.9	88.1	91.1	89.5	83.4	134.1	130.5	118.6	163.0	198
200	177.7	89.0	92.0	90.5	84.4	135.4	131.9	119.8	164.7	200
202	179.4	89.9	93.0	91.4	85.3	136.8	133.2	121.0	166.4	202
204	181.2	90.0	94.0	92.4	86.3	138.2	134.6	122.3	168.0	204
206	183.0	91.8	94.9	93.4	87.3	139.6	135.9	123.5	169.7	206
208	184.8	92.8	95.9	94.4	88.3	141.0	137.3	124.7	171.4	208
210	186.5	93.7	96.9	95.4	89.2	142.3	138.6	126.0	173.0	210
212	188.3	94.6	97.8	96.3	90.2	143.7	140.0	127.2	174.7	212
214	190.1	95.6	98.8	97.3	91.2	145.1	141.4	128.4	176.4	214
216	191.9	96.5	99.8	98.3	92.2	146.5	142.7	129.6	178.0	216
218	193.6	97.5	100.8	99.3	93.2	147.9	144.1	130.9	179.7	218
220	195.4	98.4	101.7	100.3	94.2	149.3	145.4	132.1	181.4	220
222	197.2	99.4	102.7	101.2	95.1	150.7	146.8	133.3	183.0	222
224	199.0	100.3	103.7	102.2	96.1	152.0	148.1	134.5	184.7	224
226	200.7	101.3	104.6	103.2	97.1	153.4	149.5	135.8	186.4	226
228	202.5	102.2	105.6	104.2	98.1	154.8	150.8	137.0	188.0	228
230	204.3	103.2	106.6	105.2	99.1	156.2	152.2	138.2	189.7	230
232	206.1	104.1	107.6	106.2	100.1	157.6	153.6	139.4	191.3	232
234	207.9	105.1	108.6	107.2	101.1	159.0	154.9	140.7	193.0	234
236	209.6	106.0	109.5	108.2	102.1	160.3	156.3	141.9	194.7	236
238	211.4	107.0	110.5	109.2	103.1	161.7	157.6	143.2	196.3	238
240	213.2	108.0	111.5	110.1	104.0	163.1	159.0	144.4	198.0	240
242	215.0	108.9	112.5	111.1	105.0	164.5	160.3	145.6	199.7	242
244	216.7	109.9	113.5	112.1	106.0	165.9	161.7	146.9	201.3	244
246	218.5	110.8	114.5	113.1	107.0	167.3	163.1	148.1	203.0	246
248	220.3	111.8	115.4	114.1	108.0	168.7	164.4	149.3	204.7	248

MUNSON AND WALKER'S TABLE—*Continued*

(Expressed in Milligrams)

Cuprous Oxide (Cu_2O)	Copper (Cu)	Dextrose (*d*-Glucose)	Invert Sugar	Invert Sugar and Sucrose: 0.4 gram total sugar	Invert Sugar and Sucrose: 2 grams total sugar	Lactose: $C_{12}H_{22}O_{11}+H_2O$	Lactose and Sucrose: 1 lactose, 4 sucrose	Lactose and Sucrose: 1 lactose, 12 sucrose	Maltose: $C_{12}H_{22}O_{11}+H_2O$	Cuprous Oxide (Cu_2O)
250	222.1	112.8	116.4	115.1	109.0	170.1	165.8	150.6	206.3	250
252	223.8	113.7	117.4	116.1	110.0	171.5	167.2	151.8	208.0	252
254	225.6	114.7	118.4	117.1	111.0	172.8	168.5	153.1	209.7	254
256	227.4	115.7	119.4	118.1	112.0	174.2	169.9	154.3	211.3	256
258	229.2	116.6	120.4	119.1	113.0	175.6	171.3	155.5	213.0	258
260	231.0	117.6	121.4	120.1	114.0	177.0	172.6	156.8	214.7	260
262	232.7	118.6	122.4	121.1	115.0	178.4	174.0	158.0	216.3	262
264	234.5	119.5	123.4	122.1	116.0	179.8	175.3	159.3	218.0	264
266	236.3	120.5	124.4	123.1	117.0	181.2	176.7	160.5	219.7	266
268	238.1	121.5	125.4	124.1	118.0	182.6	178.1	161.8	221.3	268
270	239.8	122.5	126.4	125.1	119.0	184.0	179.4	163.0	223.0	270
272	241.6	123.4	127.4	126.2	120.0	185.3	180.8	164.3	224.6	272
274	243.4	124.4	128.4	127.2	121.1	186.7	182.2	165.5	226.3	274
276	245.2	125.4	129.4	128.2	122.1	188.1	183.5	166.8	228.0	276
278	246.9	126.4	130.4	129.2	123.1	189.5	184.9	168.0	229.6	278
280	248.7	127.3	131.4	130.2	124.1	190.9	186.3	169.3	231.3	280
282	250.5	128.3	132.4	131.2	125.1	192.3	187.6	170.5	233.0	282
284	252.3	129.3	133.4	132.2	126.1	193.7	189.0	171.8	234.6	284
286	254.0	130.3	134.4	133.2	127.1	195.1	190.4	173.0	236.3	286
288	255.8	131.3	135.4	134.3	128.1	196.5	191.7	174.3	238.0	288
290	257.6	132.3	136.4	135.3	129.2	197.8	193.1	175.5	239.6	290
292	259.4	133.2	137.4	136.3	130.2	199.2	194.4	176.8	241.3	292
294	261.2	134.2	138.4	137.3	131.2	200.6	195.8	178.1	242.9	294
296	262.9	135.2	139.4	138.3	132.2	202.0	197.2	179.3	244.6	296
298	264.7	136.2	140.5	139.4	133.2	203.4	198.6	180.6	246.3	298
300	266.5	137.2	141.5	140.4	134.2	204.8	199.9	181.8	247.9	300
302	268.3	138.2	142.5	141.4	135.3	206.2	201.3	183.1	249.6	302
304	270.0	139.2	143.5	142.4	136.3	207.6	202.7	184.4	251.3	304
306	271.8	140.2	144.5	143.4	137.3	209.0	204.0	185.6	252.9	306
308	273.6	141.2	145.5	144.5	138.3	210.4	205.4	186.9	254.6	308
310	275.4	142.2	146.6	145.5	139.4	211.8	206.8	188.1	256.3	310
312	277.1	143.2	147.6	146.5	140.4	213.2	208.1	189.4	257.9	312
314	278.9	144.2	148.6	147.6	141.4	214.6	209.5	190.7	259.6	314
316	280.7	145.2	149.6	148.6	142.4	216.0	210.9	191.9	261.2	316
318	282.5	146.2	150.7	149.6	143.5	217.3	212.2	193.2	262.9	318
320	284.2	147.2	151.7	150.7	144.5	218.7	213.6	194.4	264.6	320
322	286.0	148.2	152.7	151.7	145.5	220.1	215.5	195.7	266.2	322
324	287.8	149.2	153.7	152.7	146.6	221.5	216.4	197.0	267.9	324
326	289.6	150.2	154.8	153.8	147.6	222.9	217.7	198.2	269.6	326
328	291.4	151.2	155.8	154.8	148.6	224.3	219.1	199.5	271.2	328

MUNSON AND WALKER'S TABLE—*Continued*

(EXPRESSED IN MILLIGRAMS)

Cuprous Oxide (Cu_2O)	Copper (Cu)	Dextrose (*d*-Glucose)	Invert Sugar	Invert Sugar and Sucrose		Lactose	Lactose and Sucrose		Maltose	Cuprous Oxide (Cu_2O)
				0.4 gram total sugar	2 grams total sugar	$C_{12}H_{22}O_{11}$ $+H_2O$	1 lactose, 4 sucrose	1 lactose, 12 sucrose	$C_{12}H_{22}O_{11}$ $+H_2O$	
330	293.1	152.2	156.8	155.8	149.7	225.7	220.5	200.8	272.9	330
332	294.9	153.2	157.9	156.9	150.7	227.1	221.8	202.0	274.6	332
334	296.7	154.2	158.9	157.9	151.7	228.5	223.2	203.3	276.2	334
336	298.5	155.2	159.9	159.0	152.8	229.9	224.6	204.6	277.9	336
338	300.2	156.3	161.0	160.0	153.8	231.3	226.0	205.9	279.5	338
340	302.0	157.3	162.0	161.0	154.8	232.7	227.4	207.1	281.2	340
342	303.8	158.3	163.1	162.1	155.9	234.1	228.7	208.4	282.9	342
344	305.6	159.3	164.1	163.1	156.9	235.5	230.1	209.7	284.5	344
346	307.3	160.3	165.1	164.2	158.0	236.9	231.5	211.0	286.2	346
348	309.1	161.4	166.2	165.2	159.0	238.3	232.9	212.2	287.9	348
350	310.9	162.4	167.2	166.3	160.1	239.7	234.3	213.5	289.5	350
352	312.7	163.4	168.3	167.3	161.1	241.1	235.6	214.8	291.2	352
354	314.4	164.4	169.3	168.4	162.2	242.5	237.0	216.1	292.8	354
356	316.2	165.4	170.4	169.4	163.2	243.9	238.4	217.3	294.5	356
358	318.0	166.5	171.4	170.5	164.3	245.3	239.8	218.6	296.2	358
360	319.8	167.5	172.5	171.5	165.3	246.7	241.2	219.9	297.8	360
362	321.6	168.5	173.5	172.6	166.4	248.1	242.5	221.2	299.5	362
364	323.3	169.6	174.6	173.7	167.4	249.5	243.9	222.5	301.2	364
366	325.1	170.6	175.6	174.7	168.5	250.9	245.3	223.7	302.8	366
368	326.9	171.6	176.7	175.8	169.5	252.3	246.7	225.0	304.5	368
370	328.7	172.7	177.7	176.8	170.6	253.7	248.1	226.3	306.1	370
372	330.4	173.7	178.8	177.9	171.6	255.1	249.5	227.6	307.8	372
374	332.2	174.7	179.8	179.0	172.7	256.5	250.9	228.9	309.5	374
376	334.0	175.8	180.9	180.0	173.7	257.9	252.2	230.2	311.1	376
378	335.8	176.8	182.0	181.1	174.8	259.3	253.6	231.5	312.8	378
380	337.5	177.9	183.0	182.1	175.9	260.7	255.0	232.8	314.5	380
382	339.3	178.9	184.1	183.2	176.9	262.1	256.4	234.1	316.1	382
384	341.1	180.0	185.2	184.3	178.0	263.5	257.8	235.4	317.8	384
386	342.9	181.0	186.2	185.4	179.1	264.9	259.2	236.6	319.4	386
388	344.6	182.0	187.3	186.4	180.1	266.5	260.5	237.9	321.1	388
390	346.4	183.1	188.4	187.5	181.2	267.7	261.9	239.2	322.8	390
392	348.2	184.1	189.4	188.6	182.3	269.1	263.3	240.5	324.4	392
394	350.0	185.2	190.5	189.7	183.3	270.5	264.7	241.8	326.1	394
396	351.8	186.2	191.6	190.7	184.4	271.9	266.1	243.1	327.7	396
398	353.5	187.3	192.7	191.8	185.5	273.3	267.5	244.4	329.4	398
400	355.3	188.4	193.7	192.9	186.5	274.7	268.9	245.7	331.1	400
402	357.1	189.4	194.8	194.0	187.6	276.1	270.3	247.0	332.7	402
404	358.9	190.5	195.9	195.0	188.7	277.5	271.7	248.3	334.4	404
406	360.6	191.5	197.0	196.1	189.8	278.9	273.0	249.6	336.0	406
408	362.4	192.6	198.1	197.2	190.8	280.3	274.4	251.0	337.7	408

MUNSON AND WALKER'S TABLE—*Continued*

(Expressed in Milligrams)

Cuprous Oxide (Cu_2O)	Copper (Cu)	Dextrose (*d*-Glucose)	Invert Sugar	Invert Sugar and Sucrose: 0.4 gram total sugar	Invert Sugar and Sucrose: 2 grams total sugar	Lactose: $C_{12}H_{22}O_{11}+H_2O$	Lactose and Sucrose: 1 lactose, 4 sucrose	Lactose and Sucrose: 1 lactose, 12 sucrose	Maltose: $C_{12}H_{22}O_{11}+H_2O$	Cuprous Oxide (Cu_2O)
410	364.2	193.7	199.1	198.3	191.9	281.7	275.8	252.3	339.4	410
412	366.0	194.7	200.2	199.4	193.0	283.2	277.2	253.6	341.0	412
414	367.7	195.8	201.3	200.5	194.1	284.6	278.6	254.9	342.7	414
416	369.5	196.8	202.4	201.6	195.2	286.0	280.0	256.2	344.4	416
418	371.3	197.9	203.5	202.6	196.2	287.4	281.4	257.5	346.0	418
420	373.1	199.0	204.6	203.7	197.3	288.8	282.8	258.8	347.7	420
422	374.8	200.1	205.7	204.8	198.4	290.2	284.2	260.1	349.3	422
424	376.6	201.1	206.7	205.9	199.5	291.6	285.6	261.4	351.0	424
426	378.4	202.2	207.8	207.0	200.6	293.0	287.0	262.7	352.7	426
428	380.2	203.3	208.9	208.1	201.7	294.4	288.4	264.0	354.3	428
430	382.0	204.4	210.1	209.2	202.7	295.8	289.8	265.4	356.0	430
432	383.7	205.5	211.1	210.3	203.8	297.2	291.2	266.6	357.6	432
434	385.5	206.5	212.2	211.4	204.9	298.6	292.6	268.0	359.3	434
436	387.3	207.6	213.3	212.5	206.0	300.0	294.0	269.3	361.0	436
438	389.1	208.7	214.4	213.6	207.1	301.4	295.4	270.6	362.6	438
440	390.8	209.8	215.5	214.7	208.2	302.8	296.8	272.0	364.3	440
442	392.6	210.9	216.6	215.8	209.3	304.2	298.2	273.3	365.9	442
444	394.4	212.0	217.8	216.9	210.4	305.6	299.6	274.6	367.6	444
446	396.2	213.1	218.9	218.0	211.5	307.0	301.0	275.9	369.3	446
448	397.9	214.1	220.0	219.1	212.6	308.4	302.4	277.2	370.9	448
450	399.7	215.2	221.1	220.2	213.7	309.9	303.8	278.6	372.6	450
452	401.5	216.3	222.2	221.4	214.8	311.3	305.2	279.9	374.2	452
454	403.3	217.4	223.3	222.5	215.9	312.7	306.6	281.2	375.9	454
456	405.1	218.5	224.4	223.6	217.0	314.1	308.0	282.5	377.6	456
458	406.8	219.6	225.5	224.7	218.1	315.5	309.4	283.9	379.2	458
460	408.6	220.7	226.7	225.8	219.2	316.9	310.8	285.2	380.9	460
462	410.4	221.8	227.8	226.9	220.3	318.3	312.2	286.5	382.5	462
464	412.2	222.9	228.9	228.1	221.4	319.7	313.6	287.8	384.2	464
466	413.9	224.0	230.0	229.2	222.5	321.1	315.0	289.2	385.9	466
468	415.7	225.1	231.2	230.3	223.7	322.5	316.4	290.5	387.5	468
470	417.5	226.2	232.3	231.4	224.8	323.9	317.7	291.8	389.2	470
472	419.3	227.4	233.4	232.5	225.9	325.3	319.1	293.2	390.8	472
474	421.0	228.3	234.5	233.7	227.0	326.8	320.5	294.5	392.5	474
476	422.8	229.6	235.7	234.8	228.1	328.2	321.9	295.8	394.2	476
478	424.6	230.7	236.8	235.9	229.2	329.6	323.3	297.1	395.8	478
480	426.4	231.8	237.9	237.1	230.3	331.0	324.7	298.5	397.5	480
482	428.1	232.9	239.1	238.2	231.5	332.4	326.1	299.8	399.1	482
484	429.9	234.1	240.2	239.3	232.6	333.8	327.5	301.1	400.8	484
486	431.7	235.2	241.4	240.5	233.7	335.2	328.9	302.5	402.4	486
488	433.5	236.3	242.5	241.6	234.8	336.6	330.3	303.8	404.1	488
490	435.3	237.4	243.6	242.7	236.0	338.0	331.7	305.1	405.8	490

SUGAR-BOILING TESTS

Use 1 quart of water to 6 pounds of sugar.

No. 1. COMMON TEST (210°–220° F.)

Dip skimmer into the boiling sugar. If the sugar falls quickly and in large drops it is ready.

No. 2. SMOOTH STAGE TEST (218°–220° F.)

Dip the tip of the forefinger into it. Press the finger and thumb together and pull quickly apart again. If it pulls to a thread which breaks readily, then the sugar is ready. The sugar has an oily feeling, therefore the name "Smooth Stage."

No. 3. LARGE THREAD TEST (227°–230° F.)

Proceed as under No. 2. The thread must not break and should be pulled so any time.

No. 4. SMALL BLOW TEST (230°–236° F.)

Dip skimmer into boiling sugar, let it drip slightly and then blow through the holes. If on the other side of the skimmer small bubbles appear, the sugar is ready.

No. 5. BLOW TEST (238°–240° F.)

Proceed as under No. 4, but when blowing through the skimmer, the bubbles must fly like soap bubbles.

No. 6. SOFT BALL TEST (242°–245° F.)

Dip your forefinger in ice water, then quickly in the boiling sugar and again in the ice water. If the sugar on finger can be rounded in a soft ball, the test is correct.

No. 7. LARGE BALL TEST (250°–255° F.)

Proceed as under No. 6, and when a large, round ball is obtained the sugar is ready.

No. 8. SMALL CRACK TEST (265°–270° F.)

Proceed as under No. 6. If the sugar breaks while in the water, but clings to the teeth if chewed, the degree is reached.

No. 9. HARD CRACK TEST (280°–310° F.)

Proceed as under No. 6. If the sugar cracks loudly and chews fine, without sticking to the teeth, it is ready.

No. 10. CARAMEL TEST (320°–350° F.)

When the sugar turns yellow and then a nice golden brown, the caramel state is reached. Do not burn the sugar, but make it as dark as wanted.

For use as coloring, add an equal amount of water to the sugar, boil up, and allow to cool. Bottle for use as coloring and flavoring.

Note: When boiling sugar, remove scum and other unclean matter which rises to the top and settles on the sides of the kettle with a skimmer. Wash down the sides of the kettle with a wet sponge to prevent the sugar from graining.

OTHER SUGAR-COOKING TESTS

Thread or Pearl	220° F. (104° C.)
Soft Ball	240° F. (116° C.)
Hard Ball	250° F. (121° C.)
Soft Crack	260° F. (127° C.)
Hard Crack	290° F. (143° C.)

ANALYSIS OF SPICES

MOISTURE

Take 1 gram of powdered spice which will pass a 60-mesh sieve and dry at 110° C. until a successive weighing shows a gain, which denotes that oxidation has begun. This takes about 12 hours or overnight. The loss is water, together with the largest part of the volatile oil. Deduction of the volatile oil, as determined in the ether extract, will give a close approximation of water.

ASH

The ash portion is determined by incineration at a very low temperature, such as may be attained in a gas muffle, which is the most convenient arrangement for work of this kind. Take 2 grams of dry material. The ash should be white or grayish-white. The proportion of ash insoluble in hydrochloric acid may be determined, when there is a reason to believe that sand is present.

SOLUBLE AND INSOLUBLE ASH

Add water to the ash in the platinum dish (or any other crucible used for the ashing), heat nearly to boiling, filter through an ashless filter, and wash with hot water until the combined filtrate and washings measure about 60 cubic centimeters. Return the filter and contents to ashing dish, ignite carefully, cool and weigh.

Calculate the per cent of water soluble and water insoluble ash.

ASH INSOLUBLE IN ACID

Boil the water-insoluble residue (or the total ash) with 25 cubic centimeters of dilute hydrochloric acid (1 : 2.5) for 5 minutes, collect the insoluble matter on an ashless filter or on a Gooch crucible. Wash with hot water, ignite, cool, and weigh.

VOLATILE AND NON-VOLATILE ETHER EXTRACT

Extract 2 grams of the ground material for 20 hours in a continuous extraction apparatus with anhydrous ether. Transfer the ethereal solution to a weighed evaporating dish and allow to evaporate at room temperature. Let stand for 18 hours over sulphuric acid and weigh as total ether extract. Heat the extract gradually and then to constant weight at 110° C. The loss is volatile ether extract. The residue is non-volatile ether extract.

CRUDE FIBER

Extract 2 grams with ether or use residue from ether extract.

Digest 2 grams of the fat-free sample with 200 cubic centimeters of a 1.25% sulphuric acid solution by boiling 30 minutes in a 1-liter Erlenmeyer flask, connected to an air condenser. Then add directly, 200 cubic centimeters of a 3.52% NaOH solution; the solution should be standardized. Boil the whole for 30 minutes, filter through linen, and wash free from alkali with hot water. Then wash thoroughly with boiling 1.25% H_2SO_4 solution, which will remove any material precipitated by the addition of the alkali. Wash free from acid. Transfer from the linen to a weighed Gooch crucible, wash with alcohol, then with ether, and dry to constant weight at 100° C. Weigh the residue, ignite, cool, and weigh. The loss is crude fiber.

A SIMPLER METHOD

Select 2 grams of the substance. Add 200 cubic centimeters of 5% hydrochloric acid, heat on a steam bath for 2 hours, raising the liquid to 90°–95° C. Filter on linen cloth, washing back into beaker with 200 cubic centimeters of 5% caustic soda solution. Heat on steam bath for 2 hours. Filter on asbestos, wash with hot water, alcohol, ether. Dry at 120° C. Weigh, ignite, cool, and weigh again. Loss in weight is crude fiber.

STARCH

Three grams of the substance are treated with about 50 cubic centimeters of cold water for 1 hour with frequent stirring. The residue is collected on a filter and washed with sufficient water to make a total of 250 cubic centimeters. This liquid contains the soluble carbohydrates. The undissolved residue is heated for 2½ hours with 2.5% hydrochloric acid (200 cubic centimeters of water and 20 cubic centimeters of hydrochloric acid, specific gravity 1.125) in a flask provided with a reflux condenser, cooled, neutralized with sodium carbonate, made up to 250 cubic centimeters, filtered, and the dextrose * determined in an aliquot portion of the filtrate. The weight of dextrose multiplied by 0.9 gives the weight of starch.

Spices like

White Pepper Ginger Nutmeg	Normally a high content of starch.
Cloves Mustard Cayenne	Normally free from starch.

TESTS FOR ADULTERATIONS IN SPICES

PEPPER

ADULTERATIONS

Wheat, buckwheat, pepper husks, ground olive stones, spent ginger.

NEUSS'S TEST

True pepper turns an intense yellow when covered with strong hydrochloric acid.

MUSTARD

ADULTERATIONS

Mustard hulls, wheat, rice, turmeric, coal-tar colors, cayenne pepper.

TESTS FOR

A. FLOUR

Two grams of mustard and 5 cubic centimeters of distilled water are boiled for 10 minutes. After cooling add a few drops of iodine. If flour is present a blue color forms. (Pure mustard contains no starch.)

B. COLOR

Pure mustard, very light dull yellow. If bright yellow, artificial color.

Add strong ammonia to the mustard: Turmeric, orange-red color.

C. CAYENNE PEPPER

Boil 1 gram of the mustard for a few minutes with alcohol, filter, evaporate to dryness. Taste the residue. If pungent, cayenne pepper.

Heat a portion of the extract and smell the fumes. Irritation of the lungs and coughing will surely follow if cayenne pepper is present.

* See Starch Analysis.

SAFFRON

Test for glycerin, oil, syrup. Test for gypsum, calcium carbonate, barium sulphate.

1. Soak saffron in water. Add a little diluted nitric acid (1 : 1) and observe with a magnifying glass. True saffron does not change, but if leaves have been added, they appear light and transparent.
2. One part saffron must color 100,000 parts water visible yellow.
3. Press between white paper. If glycerin or oil is present, a fat spot will show.
4. Shake the saffron with water. Gypsum, etc., settles to the bottom.
5. Shake with ammonia: red coloration = sandalwood; yellow coloration = saffron.
6. Shake with petrol ether: saffron = no color. If color appears, turmeric is present.
7. Ash 6.5%.

MACE

TEST FOR CHEAPER GRADE

One part of the mace is extracted with 10 parts of alcohol. Add potassium chromate solution to the extract.

If Bombay mace is present, the solution becomes red and the precipitate, which is at first yellow, becomes red on standing. True mace gives a yellow solution and precipitate and the latter does not change greatly on standing.

PAPRIKA

TEST FOR ADDED COLOR

Boil a piece of wool in a solution of the sample acidified with acetic acid. Dyes the wool red.

Carotin, the natural color in paprika, is removed from the wool by petrol ether Coal-tar dyes will remain fixed.

CAYENNE PEPPER

Test for starch as under mustard. Cayenne pepper contains no starch.

GINGER

TEST FOR EXHAUSTED GINGER

Four grams of ginger in 200 cubic centimeters of water. Shake frequently. Let stand overnight. Take 50 cubic centimeters and evaporate to dryness. Cool and weigh. Should not be under 8%.

CLOVES

Test for starch as under mustard. Cloves contain no starch.

TESTING MUSTARD FOR MAYONNAISE

(From an article appearing in the "Glass Packer," February, 1934, by Lowell B. Kilgare, Ph.D.)

TEST NO. 1. MOISTURE HOLDING POWER

The mustard most suitable for salad dressings is one which will absorb and hold a considerable amount of water. Such a mustard flour should hold at least twice its weight of water and form a heavy paste, heavy enough to mold into a ball and retain its shape.

Mustard Flour	10 g.
Water	20 cc.

Make into a paste.

After standing 1 hour, examine.

Note which makes the heaviest paste.

Dilute to 4 ounces (120 cubic centimeters). The one which made the heaviest paste remains in suspension and will not separate.

TEST NO. 2. DEVELOPMENT OF FLAVOR (Cupping Test)

The basis of the well-known "cupping test" is the action of the water upon the mustards used in the blend to produce the odor and flavor, which is desired in the product.

To ½ ounce (14.175 grams) of mustard add 1 ounce (28.35 cubic centimeters) of water. Note odor. After 1 hour re-examine. Does a bouquet or fragrance develop or does it just remain the same? Now dilute the paste with water up to 8 ounces (240 cubic centimeters). Place a cap on the jar and allow to stand for a week. A good mustard should keep perfectly during this storage period, losing little if any of its bouquet. If not, it has not been properly blended.

TEST NO. 3. KEEPING QUALITIES

One-half ounce (14.175 grams) of the flour to be tested is made into a paste, using distilled water or freshly boiled and cooled water. Dilute to 8 ounces (240 cubic centimeters). The suspension of mustard flour should show no growth of mold during a month's standing at room temperature. If mold begins to grow within a week, the test should be repeated.

ANALYSIS OF ESSENTIAL OILS (TESTS, ETC.)

Essential oils are often adulterated with alcohol, fatty oils, and cheap essential oils. The amount of alcohol added to the oil can be approximately determined by the following method:

Shake in a graduated test tube equal parts of water and oil. After clarification the amount of water will be increased by the amount of added alcohol. Or add a solution of 1 part molybdic acid in 2 parts of sulphuric acid. Shake the oil first with water, which is separated. If, after adding the molybdic acid test solution, a blue color appears, alcohol is present.

TEST FOR WATER

Milkiness in shaking with petroleum benzine.

Essential oils, adulterated with fatty oils, leave a spot on white paper which does not disappear but spreads on warming. The detection of cheap foreign oils is very difficult and best obtained by smelling and tasting.

1. Put a drop of the oil on soft filter paper and carefully warm the paper. At the end of the evaporation the odor of foreign oils is predominant. Or put 1 drop of the oil in hot water and note the odor.

2. Rub 1 drop of the oil with sugar and dissolve in plenty of water. The taste will tell the addition of foreign oils.

The specific gravity rarely gives sure results as to the purity of an oil, as the specific gravity changes with the age of the oil.

The freezing of many essential oils occurs at certain fixed temperatures, and this serves quite often as a proof of the purity of the oils.

ANALYSIS OF HONEY

The ash of genuine honey is not over 0.3%. If ash is found higher, then: (1) Test for calcium sulphate ($CaSO_4$) with barium chloride test solution. An insoluble precipitate (white), not soluble in acids indicates calcium sulphate. Starch, glucose, or invert sugar may have been added to the honey. (2) Test for chlorides with silver nitrate test solution. A white flocky precipitate, soluble in ammonia, indicates chloride of silver. Molasses may have been added to the honey.

COMMERCIAL GLUCOSE (Allen's Test)

Dextrine is present in commercial glucose, but not in pure honey.

Dilute half the honey with water and add methyl alcohol (CH_3OH) in excess, constantly stirring, until there is a permanent turbidity. If glucose is present a heavy, gummy precipitate will soon form. Genuine honey gives only a slight milkiness.

ANOTHER TEST FOR COMMERCIAL GLUCOSE

Dilute the honey with an equal volume of water, then add a few cubic centimeters of iodine solution (iodine, 1 gram; potassium iodide, 3 grams; water, 50 cubic centimeters).

In the presence of commercial glucose the solution turns red or violet.

INVERT SUGAR (Added Commercial Invert Sugar)

Dissolve 1 gram of resorcin in 100 cubic centimeters of concentrated hydrochloric acid. Introduce 10 cubic centimeters of a 50% honey solution into a test tube and add 5 cubic centimeters of ether. Shake gently and let stand for some time until the ether layer is clear.

Transfer 2 cubic centimeters of this clear ether solution to a small test tube and add a large drop of the fresh resorcin solution. Shake and note the color immediately. In the presence of commercial invert sugar the resorcin immediately assumes a cherry red color. Yellow to salmon shades have no significance.

ANALYSIS OF COFFEE

COUNTING TEST ON UNROASTED COFFEE BEANS

Bean	No. of Seeds in 50-cc. Measure
Fine Brown Java	187
Fine Mysore	198
Fine Neilgherry	203
Costa Rica	203
Good Ordinary Guatemala	207
Good Laguayra	210
Good Average Santos	213
Fine Long Berry Mocha	217
Good Ordinary Java	223
Fine Ceylon Plantation	225
Good Average Rio	236
Medium Plantation Ceylon	238
Manila	248
Ordinary Mocha	270
West African	313

TEST FOR PURITY

GROUND COFFEE

ALBERT SMITH'S TEST

Boil 10 grams of the sample in 250 cubic centimeters of water. Strain. Add basic lead acetate solution in slight excess. A precipitate forms, and when it has settled, the supernatant liquid will be colorless if the coffee is pure, but more or less colored if chicory is present.

ANALYSIS OF COCOA

ASH

The proportion of mineral ash in cocoa varies from 3.6% to 4.5%. In prepared cocoas it is considerably lower than in pure cocoa. The ash of pure cocoa is white.

TOTAL ASH

The ash is determined by the incineration of a weighed portion (about 2 grams) of the sample in a platinum dish.

WATER SOLUBLE ASH

Add water to the ash in the platinum dish, heat gently to boiling, filter, and wash with hot water until the combined filtrate and washings measure about 60 cubic centimeters. Return the filter and contents only if an ashless filter has been used, otherwise the contents alone to the platinum dish. Ignite, cool, and weigh.

The weight of insoluble ash subtracted from the weight of the total ash is the water-soluble ash. The proportion in genuine cocoa is about 50%.

FAT

Extract 2 grams of the sample with anhydrous ether (anhydrous-water free) or petroleum naphtha, best in a Soxhlet extraction apparatus. Evaporate to dryness, dry in desiccator, and weigh.

The fat is generally present in a proportion of 50% in pure chocolate liquor. (See Government Specifications.)

TEST FOR THE DETECTION OF FOREIGN FATS IN COCOA

Treat the fat with 2 parts of cold ether. Pure cocoa butter dissolves, forming a clear solution, whereas in presence of tallow or wax, a cloudy mixture is obtained.

ANALYSIS OF TEA

TOTAL ASH

Five grams of the sample are placed in a platinum dish and ignited over a Bunsen burner until complete incineration is accomplished. The dish is allowed to cool in a desiccator, and is then quickly weighed.

In genuine tea, the total ash should not be much below 5%, nor much above 6%, and it should not be magnetic.

In "faced teas," the proportion of total ash is sometimes 10%.

In "lie-tea," it may reach 30%.

In "spent-tea," it frequently falls below 3%, the ash being very rich in lime salts.

ASH INSOLUBLE IN WATER

The total ash obtained is washed in a beaker, and boiled with water for a considerable time. It is then brought upon a filter, washed, dried, ignited, and weighed.

In unadulterated tea, it rarely exceeds 3% of the sample taken.

ASH SOLUBLE IN WATER

This proportion is obtained by deducting the ash insoluble in water from the total ash.

Genuine tea contains from 3–3½% of soluble ash, or at least 50% of the total ash, whereas in exhausted tea the amount is often but 0.5%. The following formula can be used for the calculation of the per cent of spent tea:

$$E = (6 - 2S)\ 20$$

E = Spent tea

S = Per cent of soluble ash

ASH INSOLUBLE IN ACID

The ash insoluble in water is boiled with dilute hydrochloric acid and the residue separated by filtration, washed, ignited, and weighed.

In pure tea the remaining ash ranges from 0.3–0.8%. In "faced tea" or in tea adulterated by the addition of sand, etc., it may reach the proportions of 2–5%.

ANALYSIS AND TESTS FOR FLOUR

MOISTURE

Weigh about 2 grams of the sample on a weighing dish and dry for 5 hours in a drying oven at 100° C. (212° F.). Put in desiccator and weigh. Replace in the oven and heat to constant weight.

ASH

Weigh accurately into a weighed porcelain crucible about 5 grams of the sample. Burn over a Bunsen flame at a dull red heat, taking care not to fuse the ash. Weigh.

ABSORPTION

Weigh 25 grams of flour into a round bottom dish, add water from a burette, at first about 10–12 cubic centimeters. Stir the flour until a smooth dough ball has been formed. Then add slowly more water, drop by drop. Knead the dough ball between the fingers. Water is added until no more dough adheres to the hand. When this point is reached, the highest point of absorption is reached. Calculate.

A 60% absorption is called fair.

GLUTEN

Ten grams of the sample are mixed with sufficient water to make a stiff dough and allowed to stand for 1 hour. The mass is kneaded in a piece of linen in running water until the washings are clear. The fresh gluten thus obtained should have a faint yellow tinge, be tough and of such a consistency that it can be pulled out into threads. Gray and red glutens indicate inferior samples. Good wheat flour will yield from 7–18%.

GLUTEN OBTAINED FROM A MIXTURE OF

Wheat and rye flours is dark and viscous, no homogeneity.

Wheat and barley flours is dark, non-viscous, dirty reddish-brown.

Wheat and oat flours is dark yellow.

Wheat and maize flours is yellowish, not elastic.

Wheat and leguminous flours is from grayish red to green.

COLOR

The sample may be compared with one of known quality by laying out heaps of equal size, say 3 centimeters by 8 centimeters and 0.5 centimeter deep. If this is done on a colorless glass plate, the examination may be made with both white and colored background and the plate may subsequently be immersed in water (not over 35°) so that the colors produced on wetting may also be observed.

TEST TO DETECT BLEACHED FLOUR

Shake ½ ounce of flour with 2 ounces of gasoline or petrol ether. Let settle. Unbleached flour will impart to the fluid a yellow color. Bleached flour leaves the gasoline quite white.

SUBSTITUTED FLOURS (Vogel's Method)

Mix

Alcohol (70%)	95 parts
Hydrochloric Acid	5 parts

Shake flour with this reagent in a test tube. Heat to boiling. Allow to settle.

FLUID

Colorless	Pure
Straw colored tint	Presence of gruffs with bran
Orange yellow	Presence of corn cockle flour.
Flesh color	Presence of ergot.
Green	Presence of buckwheat flour

The presence of starch is detected by adding iodine solution to the boiling water solution of the flour.

Blue to black coloration, according to the amount of starch, results.

ANALYSIS OF STARCH

TEST FOR STARCH

Any solution containing starch is turned blue by adding iodine test solution.

MOISTURE

For ordinary purposes it is sufficient to obtain an exact determination of the moisture present in the starch (see baking powder).

Air-dried starch contains from 13–18% water.

Kiln-dried starch contains no water.

Kiln-dried starch absorbs readily from 7–10% moisture from the air.

DETERMINATION OF STARCH (ACID CONVERSION METHOD)

Three grams of the substance are treated with about 50 cubic centimeters of cold water for an hour, with frequent stirring. The residue is collected on a filter and washed with sufficient water to make a total of 250 cubic centimeters. This liquid contains the soluble carbohydrates. The undissolved residue is heated for 2½ hours with 2.5% hydrochloric acid (200 cubic centimeters of water and 20 cubic centimeters of HCl, specific gravity 1.125) in a flask provided with a reflux condenser, cooled, neutralized with caustic soda solution, made up to 250 cubic centimeters, filtered, and the dextrose determined in 50 cubic centimeters of the filtrate. Add 30 cubic centimeters of each of the Fehling solutions and determine the dextrose according to the method given under sugar analysis.

The weight of dextrose obtained, multiplied by 0.90, gives the weight of starch.

ANALYSIS OF LEMON AND VANILLA EXTRACT

LEMON EXTRACT

Five per cent lemon oil (as required for lemon extract) requires 80% alcohol. Fifty cubic centimeters of water in a test tube, add 2 cubic centimeters of extract. If lemon oil is present, water will be milky.

VANILLA EXTRACT (Lead Number)

A pure vanilla extract should yield with normal lead acetate solution a heavy precipitate and should settle after a few minutes, leaving a clear, supernatant, partially decolorized liquid. Mere cloudiness may be due to caramel. No precipitate indicates artificial vanilla flavor.

TEST FOR COUMARIN

When potassium iodide solution is added in excess to an aqueous solution of coumarin it forms a precipitate, at first brown and flocculent, afterwards on shaking clotting together to form a dark green, curdy mass, leaving the liquid perfectly clear.

TEST FOR CARAMEL

Dilute 10 cubic centimeters of the substance to be tested with 2 cubic centimeters of distilled water, transfer to a test tube, and shake gently for 2 minutes with 16 cubic centimeters of a mixture of:

100 cc. Amyl Alcohol 3 cc. Phosphoric Acid 3 cc. Distilled Water	Marsh Reagent

Allow the layers to separate completely. The lower aqueous layer is colorless or very nearly so, if no caramel is used. Caramel is almost insoluble in acid amyl alcohol.

The strength of an extract is denoted by the manufacturer as 1-ounce, 1½-ounce, 2-ounce goods, according to the amount required to flavor 1 gallon of simple syrup of 30° to 34° Baumé.

For testing out small quantities of syrup, accurate results are obtained by:

One ounce of extract (or 480 drops) to 1 gallon of syrup, which is equivalent to:

¼ oz.	(or 120	drops) to 1 qt. syrup
1 dr.	(or 60	drops) to 1 pt. syrup
½ dr.	(or 30	drops) to 8 oz. syrup
¼ dr.	(or 15	drops) to 4 oz. syrup
⅛ dr.	(or 7½	drops) to 2 oz. syrup

ANALYSIS OF GUMS

ANALYSIS OF GUM TRAGACANTH

Tragacanth yields a practically colorless solution when boiled with aqueous phosphoric acid. Indian gum, on the other hand, gives a pink or rose solution.

TEST FOR STARCH

Prepare a mucilage by mixing 1 gram of tragacanth with 50 cubic centimeters of water. Dilute with more water and filter. The residue is colored black-blue on the addition of iodine test solution, but not so the filtrate. If the latter shows a blue color on adding the iodine test solution, starch is present as an adulterant.

TEST FOR GUM ARABIC

To the mucilage of 1 part tragacanth and of 50 parts water add 2 grams of freshly prepared guaiac tincture (1 gram guaiac resin in 50 cubic centimeters of alcohol). No blue coloration of the liquid results after 3 hours, otherwise the tragacanth is adulterated with gum arabic.

TEST FOR INDIAN GUM

Moisten 2 grams of tragacanth in a 100-cubic centimeter beaker with a little alcohol. Add 50 cubic centimeters of distilled water, cover the beaker with the hand, and shake well. Now add 50 cubic centimeters of 4% borax solution and let stand overnight. The mucilage formed must be of such consistency to be poured out of the beaker without drawing threads. If sticky, Indian gum is present.

ANALYSIS OF BALSAMS

PERU BALSAM

On account of the high price it demands in commerce, Peru balsam is often adulterated with Canada or Copaiba balsam, castor oil, alcoholic solutions of styrax, and other resins. To test balsam of Peru for adulterations, shake an equal volume of the balsam and benzine in a graduated test tube. If pure, the volume of the benzine is not increased and the color of the benzine not discolored.

ANALYSIS OF COUMARIN AND VANILLIN

TO TEST VANILLIN FOR ADULTERATIONS

1. Take melting point. Good vanillin does not melt below 80°–81° C.
2. No residue should be left on burning on platinum foil.
3. Boil 0.2 gram vanillin with 2 cubic centimeters of concentrated HCl and add 4 cubic centimeters of phenol solution and add filtered bleaching powder solution. A dirty violet tint should not appear. Add excess of ammonia. The violet color, if present, should not become blue. (Acetanilide.)

20 grams of Vanillin = 1 kilogram of Vanilla Beans

TO TEST COUMARIN FOR ADULTERATIONS

1. Take melting point. Should melt at 67° C.
2. One-tenth gram of coumarin is boiled with 1 cubic centimeter of HCl for a minute. The clear solution is treated with 2 cubic centimeters of phenol solution and a little chloride of lime (bleaching powder). The solution should not give a red coloration. A red color, turning to an indigo blue when excess of ammonia is added, indicates the presence of acetanilide.

1 gram of Coumarin = 60 grams of Tonka Beans (60 grams = 2.12 ounces)

ANALYSIS OF CANNED FRUIT

In the case of canned fruits, an examination of the syrup in which the fruits are preserved is often sufficient.

METHOD OF DETERMINING DRAINED WEIGHT

To determine drained weight, the contents of No. 2½ cans or smaller cans should be emptied on a circular ⅛-inch mesh screen, 8 inches in diameter, set in a frame with a vertical side higher than the level of the product on the screen. The contents of the can should be distributed over the screen so as to form a layer of uniform depth, this being accomplished, so far as possible, by the manner of emptying from the can. When necessary, fruit in halves is turned over with the pit side down to permit draining of any liquid present in the pit cavity, but this should be done in such a way as to express no additional amount of liquor from the material. The period of draining should be 2 minutes. The method of determining drained weight of No. 10 cans should be the same as the foregoing, with the exception that a circular ⅛-inch mesh screen, 12 inches in diameter, be used. This screen also should be set in a frame with a vertical side higher than the level of the product on the screen.

See table for ingoing and outgoing syrup.

Use a Brix or Baumé hydrometer for sugar determination.

A Brix hydrometer gives the per cent of sugar direct. For Baumé use table.

AVERAGE COUNT AND DEGREE OF SYRUPS ON FRUIT

Items	Fancy			Ingoing Degree of Syrup	Cutout Degree of Syrup	Choice					Ingoing Degree of Syrup	Cutout Degree of Syrup	Standard		Ingoing Degree of Syrup	Cutout Degree of Syrup
	2s	2½s	10s			8 oz.	1s	2s	2½s	2s			2½s	10s		
Royal Anne Cherries		65–75	270–306	40	26	26–32	47–56		85–103	306–370	30	21	103–120	370–432	20	16
Apricots		18–24	65–86	55	28	9–11	13–16	16–20	24–30	86–108	40	22	30–36	108–130	25	16
Peaches	6–8	7–9	25–32	55	28	4–5	6–7	8–9	9–11	32–39	40	22	11–13	39–47	25	19
Melba	6–8	5–6		55	28						40	22				16
Mommoth	6–8				28				6–7			22				16
Pears		7–9	25–32	40	25		6–7	8–9	9–11	32–39	30	21	11–13	39–47	20	16
Melba		6–7		40	25							21				16
Plums																
Gage		8–10			27	5–6			11–13	39–47	40	22	16–18	47–54	25	16
Egg		8–10			27	5–6			11–13	39–47	40	22	16–18	47–54	25	16
Pineapple		8	50	50	24				8	50	35	20				
Fruit Salad				40	22						35	19				

FRUIT SALAD—Fancy......1s Tall......4 Portions. 2½s–6 Portions. 10s–19 Portions. Choice......8 ozs.–2 Portions.
TESTS for Arsenic and Heavy Chemicals—see under Canned Vegetables.

TOTAL SOLIDS

(For Apple Butter and Similar Products)

Weigh accurately 20 grams of pulped fresh fruit or a quantity of fruit products, that will give no more than 3-4 grams dry material, into a large flat-bottomed dish. If necessary to secure a thin layer of the material, add a few cubic centimeters of H_2O and mix thoroughly. Dry at 70° C. under a pressure not to exceed 100 millimeters of mercury until consecutive weighings at intervals of 2 hours do not vary more than 3 milligrams.

Read Definition of Preserve, Fruit Preserve, Jam, Fruit Jam at the end of this book.

CANNED VEGETABLES

Note carefully the external appearance of packages to detect

FLIPPERS, SPRINGERS, SWELLS, LEAKERS

Flippers and Springers are cans which may appear normal in ordinary weather but have bulged ends on hot days.

Swells. When a can is spoiled, it is usually in the condition termed "blown," i.e., with its ends convex, instead of normal or concave.

Leakers are cans which have small openings from which contents of the can escapes.

ANALYSIS OF CANNED VEGETABLES

LIQUIDS AND SOLIDS

With samples composed of solid and liquid portions proceed as follows:

1. Weigh the can—gross weight.
2. Open the can.
3. Pour the entire contents on a round sieve.

Use for No. 3 cans or smaller sizes	8-inch diameter sieve, ⅛-inch mesh screen *
Use for larger sizes	12-inch diameter sieve, ⅛-inch mesh screen *

* For canned tomatoes use a ¼-inch mesh screen.

Allow to drain 2 minutes.

4. Measure liquid. Weigh solids.
5. Weigh empty can—tare.
6. Determine net contents—gross tare.

MOISTURE

Regulate electric air oven to 135° C. Weigh 2 grams into a low dish (metal aluminum cap). Distribute contents evenly. Dry for 2 hours. Place in desiccator. Cool. Weigh. Loss in weight is moisture.

ASH

Ash about 2 grams of dry material (obtained from the above moisture determination).

TOMATO PRODUCTS (Like Tomato Puree or Tomato Sauce)

TOTAL SOLIDS

Weigh 5 grams of the sample into a flat-bottomed dish having a diameter of approximately 6 centimeters. Distribute evenly in a thin layer over the bottom of the dish, evaporate to dryness on a steam bath, and dry at 70° C. for 4 hours at a pressure not to exceed 100 millimeters.

SPECIFIC GRAVITY

Use pycnometer.

BACTERIOLOGICAL TEST

Mold, not more than 50% allowed. Use microscope, magnification of about 90. Use a slide which has been especially prepared for making mold counts (Howard cell). Spread a drop evenly over the glass slide and cover with a cover glass. Observe each field as to the presence or absence of mold filaments and note the result as positive or negative. Examine at least 50 fields, prepared from two or more mounts. Calculate the proportion of positive fields from the result of the examination of all the observed fields and report per cent of fields containing mold filaments.

YEAST AND SPORES

Maximum 125 in 1/60 cubic millimeter. Use microscope, magnification about 180.

Make a dilution of the sample with distilled water (1 : 2), or for samples like tomato sauce or paste use 10 cubic centimeters of the sample and 80 cubic centimeters of water (dilution 1 : 8). Mix well. Place a small drop upon the central disk of the Thoma-Zeiss counting cell and cover with a cover glass. After standing 10 minutes count the number of yeasts and spores on one-half of the ruled squares on the disk (counting eight blocks).

The total number thus obtained equals the number of organisms in 1/60 cubic millimeter if a dilution 1 : 2 is used. If a dilution 1 : 8 is used multiply by 3.

BACTERIA

Maximum 25 million per cubic centimeter. Use microscope, magnification about 500.

Make a dilution as directed under "Yeast and Spores." Proceed as in the previous chapter but let stand 15 minutes before counting the number of bacteria in a small area consisting of five of the small-sized squares. Move the slide to another portion of the field and count the number on another similar area. Count five such areas, preferably one from near each corner of the ruled portion of the slide and one from near the center.

Determine the average number of bacteria per area and multiply by 2,400,000 (for dilution 1 : 2) and by 7,200,000 (for dilution 1 : 8) to obtain the number of bacteria per cubic centimeter.

BIBLIOGRAPHY

U. S. Bur. Chem. Bulls. 125 and 151.
U. S. Dept. Agr. Bull. 196.
U. S. Bur. Chem. Cir. 54.
U. S. Bur. Chem. Cir. 68, p. 4.
Research Laboratory National Canners' Association Bull. 2.

TEST FOR HEAVY METALS

Cu, Copper. Zn, Zinc. Fe, Iron. Pb, Lead. Sn, Tin

Conduct evaporation in a porcelain dish. Take 100 cubic centimeters for liquids. Towards the end of the evaporation, an addition of a few cubic centimeters of strong nitric and sulphuric acid should be made. The evaporation is then carefully completed and the residue ignited at a low red heat. After cooling, the ash is moistened with HNO_3 (nitric acid), and 1 drop of H_2SO_4 (sulphuric acid) and again ignited. It is then again treated with a few drops of HNO_3, which is evaporated off cautiously, the process being stopped directly acid fumes cease to be copiously evolved. The residue is then treated with hot H_2O and the solution filtered. Then analyze according to:

Aqueous solution may contain:

Cu, Zn, Fe

Add ammonia in excess. Filter. Precipitate: Fe (Also Phosphor). Filtrate: If blue, Cu.

Divide into A and B.

A

Acidify with acetic acid. Add potassium ferrocyanide solution. Brown precipitate or coloration: Cu.

B

Heat to boiling. Add potassium ferrocyanide solution. White precipitate or turbidity: Zn.

Residue: Pb, Sn

Wash. Wash with boiling solution of ammonium acetate.

FILTER

Solution:

Acidify with acetic acid. Add potassium chromate solution. Chrome yellow precipitate: Pb.

Residue:

Ignite filter paper. Fuse in porcelain crucible with potassium cyanide. Dissolve in water. Filter. Boil insoluble residue with hydrochloric acid. Dilute. Treat clear solution with mercuric chloride solution. White silky precipitate of mercurous chloride is due to: Sn.

TEST FOR ALUMINUM

Tincture of alkanna added to neutral solutions gives a reddish-violet color with orange-yellow fluorescence.

TEST FOR BORIC ACID

A few drops of the sample or of a solution are mixed with a drop of strong hydrochloric acid and a drop of a strong alcoholic solution of turmeric, evaporated to dryness at a gentle heat, and a drop of ammonia added to the residue when cold. A dull green stain shows that boric acid is present.

ANALYSIS OF MILK

SPECIFIC GRAVITY

The specific gravity of normal milk ranges from 1.027 to 1.035. The simplest method of determining the specific gravity is by the lactometer.

FAT

To 5 cubic centimeters of milk in a small Babcock tube add an equal volume of sulphuric acid (specific gravity 1.83) and enough of a mixture of equal parts of concentrated hydrochloric acid and amyl alcohol to fill the neck of the tube. Centrifugalize 5 minutes. (If no centrifuge at hand, stopper the tube and shake.) The per cent of fat is read off directly on the tube, and is accurate to within 0.5%. If the top of the fat column is not at zero, add water and centrifugalize 1 minute. (Cow milk = 3.5% fat.)

TOTAL SOLIDS

Heat at 100° C. to constant weight about 3 grams in a tared platinum, aluminum, or tin dish of 5 centimeters diameter, with or without the addition of 15–30 grams of sand. Cool and weigh. (Cow milk = 12%.)

ASH

In a weighed dish put 20 cubic centimeters of milk. Add 6 cubic centimeters of nitric acid, evaporate to dryness and burn at a low red heat till the ash is free from carbon. The ash of normal milk is about 0.7% and faintly alkaline.

TOTAL PROTEIN

For the estimation of the total proteins (casein and albumen) the following formula may be used, which gives results close to approximation: $p = t - t/1.34$, in which p = protein, t = total solids.

MILK SUGAR

Calculated by difference.

ANALYSIS OF CREAM

Cream is to be diluted to exactly ten times its volume, the specific gravity taken, and the liquid treated as milk.

CANE SUGAR

Cane sugar may be present in milk from diluted condensed milk or from calcium saccharate added as a thickening agent. Boil about 10 cubic centimeters of the milk with 0.1 gram of resorcinol and 1 cubic centimeter of hydrochloric acid for 5 minutes. The liquid will be colored rose red if cane sugar is present.

GELATIN

To about 10 cubic centimeters of the cream add about an equal volume of water and mix thoroughly. Then add 10 cubic centimeters of Millon's test solution (dilute ten times its volume with water). Shake and, after allowing the mixture to stand 5 to 10 minutes, filter. If a considerable amount of gelatin is present, the filtrate will be somewhat turbid. To a part of the filtrate add half its volume of a saturated aqueous solution of picric acid. A yellow precipitate will be produced if gelatin is present.

TESTING POWDERED MILK

TASTE TEST

Dissolve 2¼ ounces of powdered whole milk or 1½ ounces of powdered skim milk in a pint of cold water. Place the powder on top of the water and stir until dissolved, using an egg beater. Let stand 15 minutes and taste the milk. Good milk powders will have a fresh, natural taste; others may have a foreign or sour taste.

THE GLASS TEST

Let the glasses of milk stand overnight in a cool place. Next morning examine the milk through the bottom of the glass for sediment or black specks. Pour the milk slowly out of each glass into another one. Look carefully for sediment and specks as you get near the bottom of the glass. Let the glasses of milk stand in a warm place until they sour naturally and curdle. Powder made from clean milk will have a clean sourness, not unpleasant, while that made from milk which has not been so carefully handled will develop gas and off-flavors. Powder from clean milk will yield a solid, firm curd and a clear whey. Other milk will yield a curd full of gas bubbles and a dark-colored whey.

ANALYSIS OF WATER

The most important kinds of water are distilled, soft, hard, and mineral. Distilled water is the purest.

1. Temporarily hard, containing carbonates in solution.
2. Permanently hard, containing sulphates in solution.

Calcium bicarbonate is the chief ingredient of temporarily hard water. To remove it slaked lime (calcium hydroxide, $Ca(OH)_2$) may be used to soften the water.

$$Ca(HCO_3)_2 + Ca(OH)_2 = 2CaCO_3 + 2H_2O$$

When water contains calcium sulphate it may be softened by the addition of sodium carbonate.

$$CaSO_4 + Na_2CO_3 = CaCO_3 + Na_2SO_4$$

When the carbonate and sulphate are present the water may be softened by using sodium hydroxide, which acts first upon the bicarbonate, forming sodium carbonate, and the latter in turn acts upon the calcium sulphate, forming calcium carbonate, which is insoluble and precipitates out.

According to origin, water is classified

A. Rain Water.
B. Surface, River, or Lake Water.
C. Well Water.
D. Mineral Water.

BRIEF RULES FOR TESTING WATER (Qualitative)

A. If the water reddens blue litmus paper before boiling but not afterward, it is carbonated or temporarily hard. (Calcium and magnesium bicarbonates.)

B. If the water turns red litmus paper blue after boiling or develops a red color on the addition of phenolphthalein solution, it is alkaline (permanently hard). (Calcium sulphate, magnesium sulphate.)

C. If it possesses a nauseous odor and gives a black precipitate with lead acetate solution, it is sulphurous.

D. If it possesses neither of the above properties in a marked degree and leaves a large residue upon evaporation it is saline water.

E. Organic Impurities (Animal and vegetable matter). To 1 gallon of water add ¼ grain (0.016 gram) of permanganate of potash. This will impart a purple color. If this color disappears, there is organic matter present.

If water is pure, it will not become turbid or produce a precipitate with any of the following reagents:

BARYTA WATER	If a precipitate or opaqueness appears, carbonic acid is present.
CHLORIDE OF BARIUM	A precipitate indicates sulphates.
NITRATE OF SILVER	A precipitate indicates chlorides.
OXALATE OF AMMONIA	A precipitate indicates calcium salts.

Boiling for 5 minutes and filtering when cool will in many instances make objectionable water suitable for extract purposes.

SIMPLE TEST FOR HARDNESS IN WATER

Add a weak solution of soap to water and shake. If a permanent lather forms rapidly the water is said to be soft.

If there is difficulty in producing a lather and a curd forms on the top of the water, it is said to be hard.

ANALYSIS OF YEAST

MICROSCOPICAL EXAMINATION OF YEAST

Mix a bit of the yeast in water on the glass slip till a milky fluid is formed and stir in a drop of very weak aniline dye solution, 1 gram of crystal fuchsin in 160 cubic centimeters of water having 1 cubic centimeter of alcohol. Put on the cover glass and examine under the microscope. Living, active cells resist the stain and appear colorless or nearly so, while the decayed and lifeless cells are stained.

ANALYSIS OF WHISKEY

SPIRITUS FRUMENTI

Whiskey is an alcoholic liquid obtained by the distillation of the fermented mash of wholly or partly malted cereal grains, and containing not less than 47% and not more than 53% by volume of C_2H_5OH, at 15.56° C. It must have been stored in charred-wood containers for a period of not less than 4 years.

DESCRIPTION AND PHYSICAL PROPERTIES

A light to deep amber-colored liquid, having a characteristic odor and taste, and an acid reaction.

Specific gravity from 0.935 to 0.923 at 25° C.

TESTS FOR IDENTITY AND PURITY

The residue obtained by evaporating 20 cubic centimeters of whiskey in a dish on a water bath is not completely soluble in 5 cubic centimeters of distilled water. Filter the solution and add to the filtrate a drop of diluted ferric chloride test solution (1 in 10). A greenish black coloration is produced (indicating storage in wood barrels).

A 25-cubic centimeter portion of whiskey, diluted with 50 cubic centimeters of distilled water, requires for neutralization not less than 1.5 cubic centimeters and not more than 5 cubic centimeters of tenth normal sodium hydroxide, using 3 drops of phenolphthalein test solution as indicator.

Mix 100 cubic centimeters of whiskey with 15 cubic centimeters of distilled water and slowly distill 100 cubic centimeters, using an efficient condenser. Neutralize 50 cubic centimeters of the distillate with tenth normal sodium hydroxide, using 5 drops of phenolphthalein test solution as indicator, then add exactly 20 cubic centimeters more of the tenth normal sodium hydroxide, and boil for 1 hour under a reflux condenser. Cool and titrate the excess of alkali with tenth normal sulphuric acid. Run a blank test, using distilled water in place of the distillate, and make any necessary correction. The volume of tenth normal sodium hydroxide consumed is not less than 1.7 cubic centimeters and not more than 7 cubic centimeters (esters). To 5 cubic centimeters of the distillate add 2 cubic centimeters of sodium hydroxide test solution and 5 drops of a freshly prepared aqueous solution of sodium nitroprusside (1 in 50). Then add a slight excess of acetic acid: no violet tint is produced in 1 minute (acetone).

A mixture of 0.5 cubic centimeter of the distillate with 5 cubic centimeters of distilled water meets the requirements of the test for methanol (see page 706).

Acidulate 10 cubic centimeters of whiskey with 5 drops of diluted hydrochloric acid and evaporate to 5 cubic centimeters. Dilute with distilled water to 10 cubic centimeters and filter if necessary. The addition of iodine test solution or mercuric potassium iodide test solution yields no precipitate (alkaloids).

Dilute 10 cubic centimeters of whiskey with 2 cubic centimeters of distilled water, transfer to a test tube, and shake gently for 2 minutes with 15 cubic centimeters of a mixture of 100 cubic centimeters of amyl alcohol, 3 cubic centimeters of phosphoric acid, and 3 cubic centimeters of distilled water, and allow the layers to separate completely: the lower aqueous layer is colorless or very nearly so (caramel).

Mix 5 cubic centimeters of whiskey with 5 cubic centimeters of distilled water and shake the mixture with 10 cubic centimeters of purified petroleum benzine Transfer the separated benzine to a dish of from 70 to 80 cubic centimeters capacity, add 1 cubic centimeter of sodium hydroxide solution (1 in 10), and evaporate to dryness on a water bath. Add to the dry residue 2 cubic centimeters of sulphuric acid, rotating the acid in the dish so that the residue will be completely moistened. Heat for 2 minutes on a water bath, and pour the solution into a dry test tube containing from 0.03 to 0.05 gram of resorcinol. Heat the mixture for 3 minutes at 160° to 170° C., shaking at intervals to dissolve all of the resorcinol. Then pour the solution into a mixture of 70 cubic centimeters of distilled water and 30 cubic centimeters of sodium hydroxide solution (1 in 10), adding more sodium hydroxide, if necessary, to make the mixture alkaline. The liquid shows no yellowish-green fluorescence after standing 24 hours (diethylphthalate).

Mix 2 cubic centimeters of an aqueous solution of phoroglucinol (1 in 100) with 5 cubic centimeters of sodium hydroxide test solution and add 2 cubic centimeters of whiskey: no red color is produced (formaldehyde).

Mix 20 cubic centimeters of whiskey with 20 cubic centimeters of distilled water and shake the mixture with 10 cubic centimeters of ether. Allow the mixture to stand until separation takes place, separate the ether layer, and allow it to evaporate spontaneously on a watch glass: the residue has no disagreeable or irritating odor. Evaporate 20 cubic centimeters of whiskey in a dish on a water bath and dry the residue to constant weight at 100° C.: the weight of the residue does not exceed 0.10 gram. This residue is not sticky, it has a slightly astringent taste, but is not distinctly sweet or bitter (glycerin, sugar, etc.). Evaporate 10 cubic centimeters of whiskey to 5 cubic centimeters and dilute with 10 cubic centimeters of distilled water, acidulate with 5 drops of hydrochloric acid, and add 10 cubic centimeters of hydrogen sulphide test solution: no precipitate is formed before or after rendering alkaline with ammonia test solution (heavy metals).

Place 2 cubic centimeters of mercuric sulphate test solution in a test tube, add 5 drops of whiskey and heat the mixture just to boiling over a small Bunsen burner and remove it from the flame: no yellow precipitate is formed (isopropyl alcohol).

The addition of an excess of bromine test solution to 5 cubic centimeters of whiskey, diluted with an equal volume of distilled water, produces no precipitate (phenols).

TEST FOR METHYL ALCOHOL

Mix 0.5 cubic centimeter of the distillate with 5 cubic centimeters of distilled water. Add 0.5 cubic centimeter of phosphoric acid and 2 cubic centimeters of potassium permanganate solution (3%). Allow mixture to stand for 10 minutes. Now add 1 cubic centimeter of oxalic acid solution (10%), and allow to stand until the liquid is a transparent brown. Follow with 5 cubic centimeters of diluted sulphuric acid solution (3 ounces of distilled water, 1 ounce of H_2SO_4) and 5 cubic centimeters of Schiff's reagent. Mix well and allow to stand for 10 minutes. At the end of this time, the solution, when observed against a white background, may have a reddish or pale green tint, but not a distinct blue or violet color (methanol).

TESTS FOR METHYL, ISOPROPYL, AND AMYL ALCOHOLS

One-tenth gram of vanillin is dissolved in 10 mils of alcohol in a test tube and 1 mil of pure sulphuric acid carefully run down the side of the test tube to form a layer at the bottom. By slightly rotating the tube the alcohol and acid are cautiously mixed (care is needed, otherwise the sudden rise in temperature will cause violent ebullition) and the colors formed at the area of contact and of the final mixture are noticed.

Methyl Alcohol:	Area of contact	Pale mauve
	Final mixture	Pale mauve
Ethyl Alcohol:	Area of contact	Lemon yellow
	Final mixture	Colorless
Isopropyl Alcohol:	Area of contact	Bright red towards the acid layer changing to deep blue towards the alcohol layer
	Final mixture	Prussian blue
Amyl Alcohol:	Area of contact	Dull red towards the acid layer changing to deep blue towards the alcohol layer
	Final mixture	Prussian blue A white precipitate also forms

In order to differentiate more accurately between isopropyl alcohol and amyl alcohol, 10 mils of water are added to each mixture and then shaken well. With isopropyl alcohol the mixture becomes pale blue but rapidly fades, becoming water-white. With amyl alcohol the mixture separates into two layers, the upper alcoholic one being deep grass green (permanent after 2 hours) and the lower aqueous layer water-white. The white precipitate settles to the bottom.

APPROXIMATE ESTIMATION

For the approximate estimation of methyl, isopropyl, and amyl alcohols in ethyl alcohol the quantity of sulphuric acid used was increased to 3 mils. Dilutions of the three alcohols in ethyl alcohol were made, 1 in 10, 1 in 100, 1 in 1000, and 1 in 10,000, also a control of ethyl alcohol alone, the color obtained with the latter and the solution of vanillin being a distinct yellowish green.

	1 in 10	1 in 100	1 in 1000	1 in 10,000
Methyl Alcohol	Blue green	Very faintly blue green	Yellowish green	
Isopropyl Alcohol	Blue	Pale blue	Blue green	Yellowish green
Amyl Alcohol	Deep blue	Blue	Pale blue	Very faintly blue green

VARIOUS TESTS

TEST FOR ACETIC ACID OR AN ACETATE

If a solution is to be tested for acetic acid or an acetate it is boiled with a few drops of strong sulphuric acid. If acetic acid is present, its characteristic smell is observed.

BENZOIC ACID

Make the solution alkaline with ammonia and add a few drops of ferric chloride solution. In the presence of benzoic acid, a brown coloration occurs.

CITRIC AND TARTARIC ACID

Boil with a strongly alkaline solution potassium permanganate. Tartaric acid reduces it at once. With citric acid, the green color remains.

TARTARIC ACID

If a little tartaric acid or a tartrate is heated with a few drops of resorcin solution in concentrated sulphuric acid (1 : 100) until H_2SO_4 vapors develop, the fluid acquires a fine wine-red color. The smallest trace of tartaric acid is sufficient.

SALICYLIC ACID

Its neutral solution gives with ferric chloride solution an intense violet coloration.

OXALIC ACID

On adding calcium chloride solution a precipitate occurs, which is insoluble in acetic acid.

CITRIC, SUCCINIC, AND MALIC ACIDS

On precipitating with calcium chloride solution, decomposing with dilute sulphuric acid, and boiling with potassium bichromate solution: a yellow color indicates succinic acid, a green color without odor indicates citric acid, a green color with fruity odor indicates malic acid.

TEST FOR ALDEHYDE

A solution containing an aldehyde like formaldehyde, when shaken with Schiff's reagent, turns red.

ALBUMEN

Precipitate with potassium ferro- or ferricyanide solution.

GELATIN

No precipitate with potassium ferro- or ferricyanide solution. Precipitate with concentrated tannin or picric acid solution.

CARAMEL

Dilute 10 cubic centimeters of the substance to be tested with 2 cubic centimeters of distilled water, transfer to a test tube, and shake gently for 2 minutes with 15 cubic centimeters of caramel testing solution. Allow the layers to separate completely: the lower layer is colorless or very nearly so.

COLORING MATTERS IN FOODS

WOOL DYEING TEST

Wines, fruit juices, distilled liquors, flavoring extracts, vinegars, beers, syrups, etc.

Dilute 20–200 cubic centimeters of the sample with 1–3 volumes of water and boil on the steam bath with a small piece of white woolen cloth (Nun's Veiling). When the mixture contains much alcohol, heat until most of the alcohol has been removed; in other cases, take out the wool after 5 to 15 minutes and rinse with water. Then treat the liquid with 3 or 4 drops of concentrated hydrochloric acid for each 100 cubic centimeters and warm again for 10–20 minutes with a clean piece of wool. The basic dyes go on the fiber best from neutral or faintly ammoniacal solutions and, if present, will appear on the first piece of wool. Acid colors dye from neutral solutions, but more readily from those containing free acid. If the wool takes up considerable amounts of coloring matter in either case, the presence of coal-tar dyes is indicated.

SOLUTIONS FOR MAKING THE CHEMICAL TESTS PRESCRIBED IN THIS BOOK

A. REAGENTS (To be bought).

NORMAL SOLUTIONS

(For very exact analytical work (titration), quantitative determination.)

Hanus Iodine Solution

Tenth Normal (N/10) Sodium Thiosulphate Solution ($Na_2S_2O_3$)

Fifth Normal (N/5) Potassium Hydroxide Solution (KOH)

Tenth Normal (N/10) Sulphuric Acid Solution (H_2SO_4)

Tenth Normal (N/10) Sodium Hydroxide Solution (NaOH)

(These solutions are only required in case one wants to do analytical work, not merely testing.)

ACIDS AND ALKALIES

Hydrochloric Acid (HCl), C.P. (Chemically Pure)

Sulphuric Acid (H_2SO_4), C.P. (Chemically Pure)

Nitric Acid (HNO_3), C.P. (Chemically Pure)

Acetic Acid ($C_2H_4O_2$), C.P. (Chemically Pure)

Glacial Acetic Acid ($C_2H_4O_2$) C.P. (Chemically Pure)

Phosphoric Acid (H_3PO_4) C.P. (Chemically Pure)

Carbolic Acid (C_6H_6O) C.P. (Chemically Pure)

Ammonium Hydroxide (NH_4OH) (Concentrated)

OTHERS

Bromine (Br)

Mercury (Hg)

Alcohol (Ethyl) (C_2H_5OH)

Alcohol (Methyl, Wood) (CH_3OH)

Alcohol (Amyl) ($C_5H_{11}OH$)

Ether (Anhydrous = Without Water) ($(C_2H_5)_2O$)

Ether (C.P., Chemically Pure)

Petrol Ether

Chloroform ($CHCl_3$)

Carbon Tetrachloride (CCl_4)

Carbon Bisulphite (CS_2)

Starch

Chlorinated Lime

Lime (Unslaked) (CaO)

Peroxide

Guaiac Resin

Zinc (Chemically Pure)

Tincture of Alkanna

B. REAGENTS (The compounder prepares himself)

HALPHEN TEST SOLUTION

Carbon bisulphide (Containing 1% Sulphur)	100 cc.
Amyl Alcohol	100 cc.

Mix.

FURFURAL TEST SOLUTION

Sugar	0.1 cc.
Hydrochloric Acid	10 cc.

IODINE TEST SOLUTION

Dissolve

Iodine	1 g.
Potassium Iodide	3 g.
Distilled Water	50 cc.

POTASSIUM IODIDE SOLUTION (KI) (15%)

Potassium Iodide	15 g.
Distilled Water	100 cc.

STARCH TEST SOLUTION

Starch	1 g.

is triturated with 10 cubic centimeters of cold water. Then sufficient boiling water is added with constant stirring to make 200 cubic centimeters of a thin, translucent fluid. If the solution is not translucent, boil for 3 minutes, allow to cool and filter (does not keep!). Zinc chloride added to the boiling starch solution will prevent its decomposition for a long time.

VOGEL'S TEST SOLUTION

Alcohol (70%)	95 cc.
Hydrochloric Acid	5 cc.

MILLON TEST SOLUTION

Dissolve mercury in twice its weight of nitric acid. Dilute with equal volume of distilled water. To protein solution added: white precipitate. Heated: brick red. (Salt (NaCl) must not be present).

FEHLING'S SOLUTION

SOLUTION A

Crystal Copper Sulphate ($CuSO_4$)	34.639 g.

Dissolved in distilled water and made up to exactly 500 cubic centimeters.

SOLUTION B

Rochelle Salt (Potassium Sodium Tartrate, $C_4H_4O_6KNa$)	173 g.
Caustic Soda (NaOH)	50 g.

Dissolved in water and made up to exactly 500 cubic centimeters.

For use, equal quantities of the two solutions should be mixed at the time required.

MAYER'S REAGENT

Mercuric Chloride (Sublimate) ($HgCl_2$)	13.55 g.
Potassium Iodide	50.0 g.
Distilled Water	1000 cc.

This test solution gives, with most alkaloids, in a slightly acid solution: a white precipitate.

TANNIN SOLUTION (20%) (Saturated) ($C_{14}H_{10}O_9$)

Tannin	20 g.
Distilled Water	100 cc.

BORAX SOLUTION ($Na_2B_4O_7$) (4%)

Borax	4 g.
H_2O *	96 cc.

* In making up these solutions use distilled water only.

BARYTA WATER

Barium Hydroxide ($Ba(OH)_2$)	45 g.
H_2O	1000 cc.

Let stand, decant, and filter.

BARIUM CHLORIDE SOLUTION ($BaCl_2$) (10%)

Barium Chloride	10 g.
H_2O	90 cc.

Gives with sulphates: white precipitate.

CAUSTIC SODA SOLUTION (NaOH) (10%)

Sodium Hydroxide	10 g.
H_2O	90 cc.

Solutions to be tested for ammonia give up, on warming with this reagent, the characteristic odor of ammonia.

CALCIUM CHLORIDE SOLUTION ($CaCl_2$) (10%)

Calcium Chloride	10 g.
H_2O	90 cc.

FERRIC CHLORIDE SOLUTION ($FeCl_3$) (5%)

Ferric Chloride	5 g.
H_2O	95 cc.

LEAD ACETATE SOLUTION ($Pb(C_2H_3O_2)_2$) (10%)

Lead Acetate	10 g.
H_2O	90 cc.

Solution should be neutral. If alkaline, add drops of acetic acid until neutral.

LEAD ACETATE SOLUTION BASIC ($Pb(C_2H_3O_2)_2 + PbO$) (10%)

Basic Lead Acetate	10 g.
H_2O	90 cc.

MERCURIC CHLORIDE SOLUTION ($HgCl_2$) (5%)

Sublimate (Mercuric Chloride)	5 g.
H_2O	95 cc.

PHENOL SOLUTION (C_6H_5OH) (5%)

Carbolic Acid Crystals	5 g.
H_2O	95 cc.

POTASSIUM HYDROXIDE SOLUTION (KOH) (50%)

Potassium Hydroxide	50 g.
H_2O	50 cc.

POTASSIUM CHROMATE SOLUTION (K_2CrO_4) (10%)

Potassium Chromate	10 g.
H_2O	90 cc.

POTASSIUM BICHROMATE SOLUTION ($K_2Cr_2O_7$) (10%)

Potassium Bichromate	10 g.
H_2O	90 cc.

POTASSIUM FERROCYANIDE ($K_4Fe(CN)_6$) (5% Solution)

Potassium Ferrocyanide (Yellow)	5 g.
H_2O	95 cc.

With ferric compounds: blue precipitate (Prussian blue).

POTASSIUM FERRICYANIDE ($K_3Fe(CN)_6$) (5% Solution)

Potassium Ferricyanide (Red)	5 g.
H_2O	95 cc.

With ferrous compounds: blue precipitate (Turnbull blue).

POTASSIUM PERMANGANATE SOLUTION ($KMnO_4$) (3%)

Potassium Permanganate	3 g.
H_2O	97 cc.

RESORCIN-HCl SOLUTION ($C_6H_4(OH)_2$)

Resorcin	1 g.
Hydrochloric Acid	100 cc.

OXALIC ACID SOLUTION ($C_2H_2O_4$) (10%)

Oxalic Acid	10 g.
H_2O	90 cc.

SATURATED SALT SOLUTION (Sodium Chloride, NaCl) (30%)

Sodium Chloride	2 lb.
H_2O	1 gal.

SOAP SOLUTION (2%)

Castile Soap	2 g.
H_2O	98 cc.

SILVER NITRATE SOLUTION ($AgNO_3$) (5%)

Silver Nitrate	5 g.
H_2O	95 cc.

This reagent gives: with chlorides, a white, flocky precipitate; with sulfides, a black precipitate; with sulfates, none; with phosphates, a yellow precipitate.

SODIUM CARBONATE SOLUTION (Na_2CO_3) (10%)

Sodium Carbonate	10 g.
H_2O	90 cc.

SODIUM ACETATE SOLUTION ($C_2H_3O_2Na$) (10%)

Sodium Acetate	10 g.
H_2O	90 cc.

SULPHOCYANIDE OF POTASSIUM SOLUTION (KCSN) (10%)

Potassium Sulphocyanide	10 g.
H_2O	90 cc.

OXALATE OF AMMONIA SOLUTION ($C_2O_4(NH_4)_2$) (10%)

Ammonium Oxalate	10 g.
H_2O	90 cc.

This reagent gives with calcium compounds a white precipitate.

AMMONIUM CHLORIDE SOLUTION (Sal Ammoniac) (NH_4Cl) (10%)

Ammonium Chloride	10 g.
H_2O	90 cc.

COBALTOUS NITRATE SOLUTION ($Co(NO_3)_2$) (5%)

Cobaltous Nitrate	5 g.
H_2O	95 cc.

URANIUM NITRATE SOLUTION ($UO_2(NO_3)$) (2%)

Uranium Nitrate	2 g.
H_2O	98 cc.

Uranium nitrate is very poisonous.

HYDROCHLORIC ACID SOLUTION (1 : 2.5)

Hydrochloric Acid	100 cc.
Distilled Water	250 cc.

PICRIC ACID SOLUTION ($C_6H_2(NO_2)_3OH$) (10%)

Picric Acid	10 g.
H_2O	90 cc.

AMMONIUM ACETATE SOLUTION ($(NH_4)C_2H_3O_2$) (10%)

Ammonium Acetate	10 g.
H_2O	90 cc.

PHENOLPHTHALEIN INDICATOR

Dissolve

Phenolphthalein	0.1 g.
in	
Alcohol	60 cc.
and dilute with	
Water to	100 cc.

Gives with alkaline solutions a red color. It is not reliable for phosphate, bicarbonates, ammonia.

METHYL ORANGE INDICATOR

Dissolve

Methyl Orange	0.1 g.
in	
Water	500 cc.

Gives with alkaline solutions a yellow color. Gives with acid solutions an orange to yellow-red. Not to be used for organic acids.

ROSOLIC ACID INDICATOR

Dissolve

Rosolic Acid	0.2 g.
in	
Diluted Alcohol	100 cc.

Gives with alkaline solutions a red color. Gives with acid solutions a weak yellow color. Suitable for ammonia.

FUCHSIN SULPHUROUS ACID TEST SOLUTION (SCHIFF'S REAGENT)

To a solution of 0.5 gram of basic fuchsin and 9 grams of sodium bisulfide in 500 cubic centimeters of distilled water, add 10 cubic centimeters of hydrochloric acid. Preserve the solution in a well-stoppered bottle, protected from light.

CARAMEL TESTING SOLUTION (MARSH REAGENT)

Amyl Alcohol	100 cc.
Phosphoric Acid (Sirupy Phosphoric Acid)	3 cc.
Distilled Water	3 cc.

WEIGHTS AND MEASURES

COMMON KITCHEN MEASURES

Equivalents of Capacity
(All measures level full)

3 teaspoons; ½ fluid ounce	= 1 tablespoon
16 tablespoons; 2 gills; ½ liquid pint; 8 fluid ounces	= 1 cup
1 liquid pint; 16 fluid ounces	= 2 cups

DRY MEASURE

(For fruits, vegetables, and other dry commodities)

2 pints	= 1 quart
8 quarts	= 1 peck
4 pecks	= 1 bushel
(2150.42 cubic inches)	
105 quarts	= 1 barrel
(7056 cubic inches)	

LIQUID MEASURE

4 fluid ounces	= 1 gill
4 gills	= 1 pint
2 pints	= 1 quart
4 quarts	= 1 gallon
(231 cubic inches)	
31½ gallons	= 1 barrel
2 barrels	= 1 hogshead

LINEAR MEASURE

12 inches	= 1 foot
3 feet	= 1 yard
5½ yards	= 1 rod
40 rods	= 1 furlong
8 furlongs; 1760 yards; 5280 feet	= 1 mile

CUBIC MEASURE

1728 cubic inches	= 1 cubic foot
27 cubic feet	= 1 cubic yard
144 cubic inches	= 1 board foot
128 cubic feet	= 1 cord*

* The use in trade of the so-called "face cord," which is less than the above, is illegal.

SQUARE MEASURE

144 square inches	= 1 square foot
9 square feet	= 1 square yard
30¼ square yards	= 1 square rod
160 square rods	= 1 acre
640 acres	= 1 square mile
36 square miles	= 1 township

AVOIRDUPOIS WEIGHT

27 11/32 grains	= 1 dram
16 drams	= 1 ounce
16 ounces	= 1 pound
4 quarters	= 1 hundredweight
20 hundredweights	= 1 ton

25 pounds	= 1 short quarter
28 pounds	= 1 long quarter
100 pounds	= 1 short hundredweight
112 pounds	= 1 long hundredweight
2000 pounds	= 1 short ton
2240 pounds	= 1 long ton.

COMMON RULES OF MEASUREMENT

Rectangle Area = length × width

Solid with rectangular sides Volume = length × width × height

Circle
- Circumference = 3.1416 × diameter
- Area = 0.7854 × diameter × diameter

Cylinder
- Area (exclusive of that of ends) = 3.1416 × diameter × height
- Volume = 0.7854 × diameter × diameter × height

INTERNATIONAL METRIC SYSTEM

In the international metric system the fundamental unit is the meter—the unit of length. From this the units of capacity (liter) and of weight (gram) are derived. All other units are the decimal subdivisions or multiples of these. These three units are simply related; e.g., for all practical purposes 1 cubic decimeter equals 1 liter and 1 liter of water weighs 1 kilogram. The metric tables are formed by combining the words "meter," "gram," and "liter" with the six numerical prefixes, as in the following tables:

One meter = 39.37 inches; 1 liter = 1.0567 liquid quarts; 1 gram = 0.035 avoirdupois ounce.

UNITS OF LENGTH

Millimeter	=	0.001	meter
Centimeter	=	.01	meter
Decimeter	=	.1	meter
METER	=	1	meter
Decameter	=	10	meters
Hectometer	=	100	meters
Kilometer	=	1000	meters

UNITS OF CAPACITY

Milliliter	=	0.001	liter
Centiliter	=	.01	liter
Deciliter	=	.1	liter
LITER	=	1	liter
Decaliter	=	10	liters
Hectoliter	=	100	liters
Kiloliter	=	1000	liters

UNITS OF WEIGHT (OR MASS)

Milligram	=	0.001	gram
Centigram	=	.01	gram
Decigram	=	.1	gram
GRAM	=	1	gram
Decagram	=	10	grams
Hectogram	=	100	grams
Kilogram	=	1000	grams

UNITS OF AREA

The table of areas is formed by squaring the length measures, as in our common system. For land measure 10 meters square is called an "ARE" (meaning "area"). The side of 1 are is about 33 feet. The hectare is 100 meters square, and, as its name indicates, is 100 ares, or about 2½ acres.

APPROXIMATE WEIGHTS OF SOME COMMON DRY COMMODITIES

	Pounds per Bushel
Apples	48
Beans	60
Beets	60
Carrots	50
Cranberries	32
Cucumbers	48
Onions	57
Parsnips	50
Peaches	48
Peanuts	22
Pears	58
Peas (Dried)	60
Peas (Green, Unshelled)	56
Potatoes (White)	60
Potatoes (Sweet)	54
Tomatoes	56
Turnips	55

APPROXIMATE WEIGHTS OF SOME COMMON DRY COMMODITIES—(*Continued*)

Ice: 1 cubic foot	= 57.2 lb.
30 cubic inches	= 1 lb.
Sugar, granulated: 1 cup	= ½ lb.
Butter: 1 cup	= ½ lb.
Lard: 1 cup	= ½ lb.
Flour: 1 cup	= ¼ lb.
Rice: 1 cup	= ½ lb.
Cornmeal: 1 cup	= 5 oz.
Raisins (stemmed): 1 cup	= 6 oz.
Currants (cleaned): 1 cup	= 6 oz.
Bread crumbs (stale): 1 cup	= 2 oz.
Chopped meat (packed): 1 cup	= ½ lb.

These weights are approximate only and should therefore not be used to determine whether correct measure is given or received.

APPROXIMATE WEIGHTS OF SOME COMMON MATERIALS

	Pounds per Cubic Foot
Coal	
Bituminous (piled loose)	44 to 54
Anthracite (piled loose)	50 to 57
Coke (piled loose)	23 to 32
Charcoal of pine and oak	15 to 30
Earth (common loam)	
Dry (loose)	72 to 80
Moist (moderately packed)	90 to 100
Soft mud (packed)	110 to 120
Masonry	
Brickwork	100 to 140
Roughly scabbled mortar rubble	140 to 150
Mortar (hardened)	103
Sand	90 to 117
Gravel	90 to 117
Trap (quarried, in loose piles)	107

	Pounds per 1000 Board Feet	Pounds per Cubic Foot
Woods		
Fir (Balsam)	2170	26
Hemlock	2330 to 2580	28 to 31
Maple	3250 to 3920	39 to 47
Oak	3080 to 4670	37 to 56
Pine (American White)	1830 to 2580	22 to 31
Pine (Yellow)	1920 to 3080	23 to 37
Poplar	1830 to 2580	22 to 31

The wood is supposed to be seasoned and of average dryness.

BOILING POINT AND SPECIFIC GRAVITY OF VARIOUS LIQUIDS

	Boiling Point °C.	Specific Gravity 20/4°C.	Approx. Lbs. Per U.S. Gal. Room Temp.
Acetic Acid, Glacial	118.1	1.049	8.74
Acetone	56.5	.792	6.60
Acetylene Tetrachloride	146.3	1.600	13.33
Alcohol, n-Amyl	137.9	.817 (20/20° C.)	6.81
Alcohol, Butyl	117	.810	6.75
Alcohol, Ethyl, 190 Proof	78.5	.815	6.79
Alcohol, Isopropyl	82.5	.789	6.55
Alcohol, Methyl 99.5%	64.7	.797 (15/15° C.)	6.64
Amyl Acetate (Banana Oil)	142	.876 (15/4° C.)	7.30
Aniline	184.4	1.022	8.51
Benzaldehyde	179	1.046	8.71
Benzene, C.P.	79.6	.879	7.32
Benzol, 90%	—	.876–.888	7.35
Carbon Disulfide	46.3	1.263	10.52
Carbon Tetrachloride	76.8	1.595	13.29
Castor Oil	—	.960–.967 (15/15° C.)	8.02
Chlorobenzene (mono)	132.1	1.107	9.22
Chloroform	61.2	1.489	12.40
Corn Oil	—	.921–.928 (15/15° C.)	7.71
Cyclohexanol (Hexalin)	160–1	.962	8.01
p-Cymene	176–7	.857	7.14
Diethylamine	55.5	.712 (15/15° C.)	5.93
Diethyl Phthalate	298–9	1.121 (25/25° C.)	9.34
Dimethylaniline	193	.956	7.96
Ethanolamine (mono)	171	1.022	8.51
Ether	34.6	.708 (25/4° C.)	5.90
Ethyl Acetate	77.1	.901	7.51
Ethylene Dichloride	83.7	1.257	10.47
Ethylene Glycol	197.4	1.113	0.27
Gasoline 60°	—	.740	6.16
Gasoline 70°	40–160	.700	5.83
Glycerol	290	1.260	10.50
Kerosene	150–300	.820	6.83
Linseed Oil	—	.934 (15/15° C.)	7.78
Nitrobenzene	210.9	1.205 (18/4° C.)	10.04
Paraffin Oil	—	.900	7.50
Pine Oil	150–185	.870	7.25
Piperidine	106	.860	7.16
Pyridine	115–6	.982	8.18
Solvent Naphtha	125–180	.861–.869	7.21
Sulfur Chloride (mono)	138	1.687	14.05
Toluene	110.8	.866	7.21
Triethanolamine	277–9 (150 mm.)	1.124 (20° C.)	9.36
Turpentine, Oil of	155–180	.873	7.27
Water	100	1.000 (4° C.)	8.33
Xylene Comm.	135–145	.860–.870	7.21
Mercury	356.9	13.546	112.84

Abridged Edition, Price $1.00

9 780820 601533

Printed by Libri Plureos GmbH in Hamburg,
Germany